Irrigation: Systems, Technology and Impacts

Irrigation: Systems, Technology and Impacts

Lucias Chapman

⬠SYRAWOOD
PUBLISHING HOUSE

New York

Published by Syrawood Publishing House,
750 Third Avenue, 9th Floor,
New York, NY 10017, USA
www.syrawoodpublishinghouse.com

Irrigation: Systems, Technology and Impacts
Lucias Chapman

International Standard Book Number: 978-1-68286-815-7 (Hardback)

Cataloging-in-Publication Data

Irrigation : systems, technology and impacts / Lucias Chapman.
   p. cm.
Includes bibliographical references and index.
ISBN 978-1-68286-815-7
1. Irrigation. 2. Water in agriculture. 3. Irrigation engineering.
4. Irrigation farming. I. Chapman, Lucias.
S613 .I77 2019
631.587--dc23

TABLE OF CONTENTS

PREFACE

An important requirement in agriculture is controlled and regulated irrigation. It refers to the application of water in requisite amounts to crops at regular intervals. This facilitates the growth of crops, maintenance of landscapes and revegetation in disturbed soils in dry climates or areas with low rainfall. It also plays a significant role in suppressing weed growth, frost protection, dust suppression and preventing soil consolidation. The varied methods of irrigation include micro-irrigation, surface irrigation, sprinkler irrigation, center pivot irrigation, etc. Most of the topics introduced in this book cover new techniques and the applications of irrigation systems and technologies. Some of the diverse topics covered in this book address the varied types that fall under this category. This book is an essential guide for both academicians and those who wish to pursue this discipline further.

A foreword of all Chapters of the book is provided below:

Chapter 1- The application of controlled quantity of water at regular interval to crops in agriculture is called irrigation. It plays a crucial role in the maintenance of landscapes and revegetation of disturbed soils in arid areas. The aim of this chapter is to provide an introduction to irrigation through a study of water resources, desalination, sources of fresh water and importance of water resources;

Chapter 2- There are many methods of irrigation, such as surface irrigation, sprinkler irrigation, subirrigation and micro-irrigation. The topics elucidated in this chapter cover some of these important types of irrigation such as surface, drip, sprinkler, center pivot irrigation, etc.;

Chapter 3- Irrigation is a central feature in agriculture. It supports crop production, suppresses weed growth and prevents soil consolidation. A well-regulated irrigation practice is therefore of the utmost necessity in agriculture. This chapter closely examines some of the crucial aspects of irrigation in agriculture. It includes topics like agricultural water, irrigation scheduling, ditch irrigation, overhead irrigation, etc.;

Chapter 4 - The practice of growing plants with similar water requirements as a strategy of conserving water is known as hydrozoning. An elaborate study of the varied principles of hydrozoning has been provided in this chapter, which includes hydrozoning areas and benefits of hydrozones, among others;

Chapter 5- Irrigation can have significant environmental impacts. These are manifested as a change in the quality and quantity of soil and water. It also results in alteration of the social and natural conditions downstream of an irrigation scheme. This chapter discusses the environmental impacts of irrigation, which includes direct and indirect effects, reduced river flow and increased groundwater recharge.;

Chapter 6- The study of irrigation technology is vital for a holistic understanding of irrigation. This chapter closely examines the different aspects and uses of valley variable rate irrigation, site-specific management, GPS guidance, T-L irrigation, rain sensor, etc. in agriculture;

Chapter 7- There are different kinds of irrigation systems, such as manual, gravity, lift, sewage irrigation systems, etc. This chapter delves into the varied aspects of these systems to provide a holistic understanding of irrigation.

I would like to thank the entire editorial team who made sincere efforts for this book and my family who supported me in my efforts of working on this book. I take this opportunity to thank all those who have been a guiding force throughout my life.

Lucias Chapman

Chapter 1

Introduction to Irrigation

The application of controlled quantity of water at regular interval to crops in agriculture is called irrigation. It plays a crucial role in the maintenance of landscapes and revegetation of disturbed soils in arid areas. The aim of this chapter is to provide an introduction to irrigation through a study of water resources, desalination, sources of fresh water and importance of water resources.

Water Resources

Water resource is any of the entire range of natural waters that occur on the Earth, regardless of their state (i.e., vapor, liquid, or solid) and that are of potential use to humans. Of these, the resources most available for use are the waters of the oceans, rivers, and lakes; other available water resources include groundwater and deep subsurface waters and glaciers and permanent snowfields.

Water Resources of Asia

Asia's water resources constitute a vast potential, both for generating hydroelectricity and for irrigating crops. Water is important for irrigation in many Asian regions that are either arid (as in much of Central and Southwest Asia).

Human use of natural waters, particularly of freshwater resources, has increased steadily over the centuries. It is unlikely that this trend will change given the continued growth of population and the ever-widening utilization of water for agricultural, industrial, and recreational purposes. This situation has given rise to growing concern over the availability of adequate water supplies to accommodate the future needs of society. Surface-water resources are already being used to their maximum capacity in various regions of the world, as, for example, in the southwestern United States.

Quantity of water is not the only concern. Overuse has resulted in the progressive deterioration of water quality. Seepage of mineral fertilizers (phosphates and nitrates), pesticides, and herbicides into surface and subsurface waters has not only rendered them unfit for human consumption but also disrupted aquatic ecosystems. Lakes and rivers also have been contaminated by the improper disposal of sewage, the discharge of untreated industrial wastes (including such toxicants as polychlorinated biphenyls, or PCBs), and the release of heated wastewater from nuclear-power plants and other industrial facilities, which results in thermal pollution and its attendant problems.

Efforts are being made to curb the contamination of water resources. For example, regulatory action by the U.S. government to reduce phosphorous input into the Great Lakes has had measurable results, as has the implementation of improved waste-purification technology by certain municipalities in the region. The latter not only helps to restore water resources but also conserves the water supply by effective recycling. Advanced sewage-treatment facilities have made it possible to obtain potable water purer than most stream water. Projects to remove salt and other dissolved solids from brackish surface water as well as from seawater have been undertaken in such countries as Australia, Kuwait, and the United States. Water from desalination plants is generally suitable for household use and for irrigation. Other procedures employed for relieving water shortages include control of runoff and the reduction of evaporation by means of agricultural-engineering measures.

Sources of Fresh Water

Surface Water is a term used to describe water on the earth's surface, including rivers, creeks and wetlands.

Most surface water comes from rainfall (precipitation) runoff from the surrounding land area (catchment). Of course not all runoff ends up in rivers, some evaporates, some is used by vegetation and part of it soaks into the ground recharging our ground-water systems, some of which can then seep back into the riverbeds.

At a certain depth below the land surface, called the water table the ground becomes saturated with water. If a river happens to cut into this saturated layer, then water will seep out of the ground into the river. Groundwater seepage is most commonly seen in the form of springs e.g. Berry Springs, Katherine Hot Springs and Bitter Springs.

There are three types of surface water:

- Permanent (perennial) - permanent surface waters are present throughout the year. They are usually in the form of rivers, lakes, springs and swamps. At times when there is little or no rain, the water level is maintained by groundwater contributions.

- Semi-permanent (ephemeral) - Semi-permanent water bodies are those that only hold water for part of the year. These are usually small creeks, lagoons, waterholes, or low-lying areas in the arid zone.

- Man-made – surface water can also be held in manmade structures ranging from lakes, dams and turkey nests to artificial swamps and sewage treatment ponds.

The majority of surface water in the northern territory is found in the top end. This reflects the tropical climate that can bring vast quantities of rain and high humidity. The seasons (dry and wet) cause large variations in surface flow while creeks record little or no flow during the dry season and flows up to ten times the monthly average may be recorded during the wet season. Surface waters in the Arid Zone are minimal as a result of much lower annual rainfall. Waterways (clay pans, salt pans, rivers and flood outs) are dry for the vast majority of time. Infrequent but sometimes intense rainfall events cause arid zone streams to flow for short periods. Surface waters have been essential not only to humans, but to all life on earth, ever since life began. Plants and animals grow and congregate around waterways simply because water is so essential to all life. It might seem that rivers happen to run through many cities in the world, but it is not that the rivers go through the city, but rather that the city was built and grew around the river.

Groundwater

Groundwater is contained in what are called 'aquifers'. An aquifer is a geological formation or part of it, consisting of permeable material capable to store/yield significant quantities of water. Aquifers can consist of different materials: unconsolidated sands and gravels, permeable sedimentary rocks such as sandstones or limestones, fractured volcanic and crystalline rocks, etc.

Groundwater is (naturally) recharged by rain water and snowmelt or from water that leaks through the bottom of some lakes and rivers. Groundwater also can be recharged when water supply systems leak and when crops are irrigated with more water than required. There are also techniques to manage aquifer recharge and increase the amount of water infiltrating into the ground.

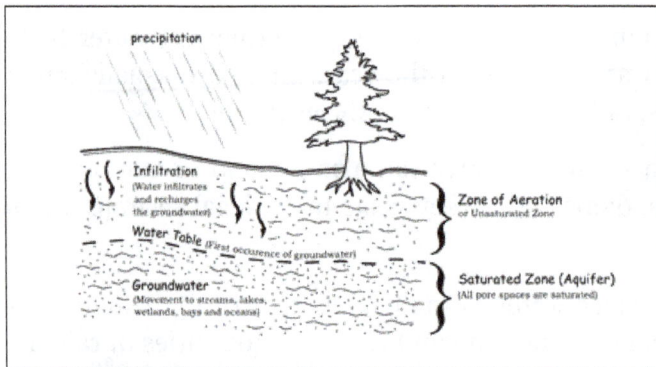

Water table, saturated and unsaturated zones

Groundwater can be found almost everywhere. The water table may lie deep or shallow depending on several factors such as the physical characteristics of the region, the meteorological conditions and the recharge and exploitation rates. Heavy rains may increase recharge and cause the water table to rise. But in the other hand, an extended period of dry weather may cause the water table to fall.

When groundwater reaches an aquifer it does not stand still. It normally will keep flowing but much slower than before reaching the aquifer. How fast groundwater flows depends on the characteristics of the aquifer. The direction it moves is normally from high to lower levels ruled by gravity, unless there is any anthropogenic impact such as pumping wells. The groundwater will move until it discharges into another aquifer or another water body like a lake, a river, the ocean or until it is extracted by a well.

To be able to store and yield groundwater, an aquifer needs to have certain physical characteristics. It needs to have empty space (pores or fractures) where groundwater can be stored and the spaces need to be connected to allow it to flow through. Technically speaking, when there are spaces and they are interconnected the geological formation is permeable. When there are no spaces or they are not interconnected, the geological formation is impermeable. The greater the aquifer's porosity and permeability are, the more groundwater is stored and yield by an aquifer.

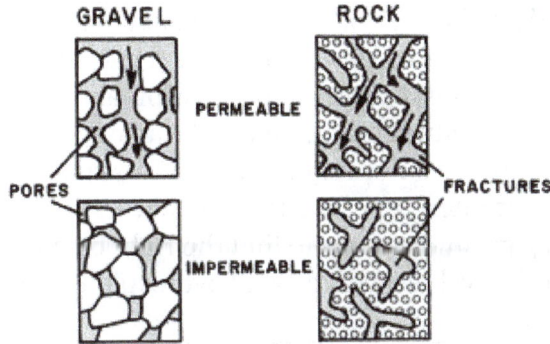

Permeable and impermeable geological formations

Importance of Groundwater

Groundwater represents about 30% of world's fresh water. From the other 70%, nearly 69% is captured in the ice caps and mountain snow/glaciers and merely 1% is found in river and lakes. Groundwater counts in average for one third of the fresh water consumed by humans, but at some parts of the world, this percentage can reach up to 100%. In the figure bellow an overview is given of Earth's water distribution.

Groundwater is a very important natural resource and has a significant role in the economy. It is the main source of water for irrigation and the food industry. In general groundwater is a reliable source of water for the agriculture and can be used in a flexible manner: when it's dry and there is larger demand more groundwater can be extracted and when the rain fall meets the necessities, less groundwater will need to be extracted. Globally, irrigation accounts for more than 70% of total water withdraw (both surface and groundwater). Groundwater is estimated to be used for circa 43% of the total irrigation water use.

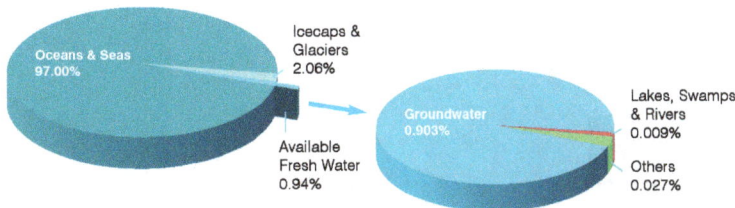

For the environment groundwater plays a very important role in keeping the water level and flow into rivers, lakes and wetlands. Specially during the drier months when there is little direct recharge from rainfall, it provides the environment with groundwater flow through the bottom of these water bodies and becomes essential for the wild life and plants living in these environment. Groundwater also plays a very relevant role in sustain navigation through inland waters in the drier seasons. By discharging groundwater into the rivers it helps keeping the water levels higher.

Groundwater is found almost everywhere and its quality is usually very good. The fact that groundwater is stored in the layers beneath the surface, and sometime at very high depths, helps protecting it from contamination and preserve its quality. Additionally, groundwater is a natural resource which can often be found close to the final consumers and therefore does not require large investments in terms of infrastructure and treatment, as it often is necessary when harvesting surface water. The most important about using groundwater is to find the right balance between withdrawing and letting the aquifer's level recover to avoid overexploitation and to avoid pollution of this crucial resource.

Frozen Water

Of the 3 percent of earth's water considered freshwater, 70 percent of that small amount is currently locked in glaciers and ice caps. In theory, frozen glacial and ice cap water could be melted and used, but the amount of energy needed to melt and transport vast quantities of ice make it economically impractical. Glaciers and ice caps also play vitally important roles in the regulation of earth's climates and global temperatures, making their preservation very important.

Desalination

Water desalination processes separate dissolved salts and other minerals from water. Feed water sources may include brackish, seawater, wells, surface (rivers and streams), wastewater, and industrial feed and process waters. Membrane separation requires driving forces including pressure (applied and vapor), electric potential, and concentration to overcome natural osmotic pressures and effectively force water through membrane processes. As such, the technology is energy intensive and research is continually evolving to improve efficiency and reduce energy consumption.

Seawater desalination has the potential to reliably produce enough potable water to support large populations located near the coast. Numerous membrane filtration seawater desalination plants are currently under construction or in the planning stages up and down California's parched coast, with the 50 million gallons per day (mgd) Carlsbad Desalination plant scheduled to be operational by 2016.

Reverse osmosis (RO) and nano filtration (NF) are the leading pressure driven membrane processes. Membrane configurations include spiral wound, hollow fiber, and sheet with spiral being the most widely used. Contemporary membranes are primarily polymeric materials with cellulose acetate still used to a much lesser degree. Operating pressures for RO and NF are in the range of 50 to 1,000 psig (3.4 to 68 bar, 345 to 6896 kPa).

Electrodialysis (ED) and electrodialysis reversal (EDR) processes are driven by direct current (DC) in which ions (as opposed to water in pressure driven processes) flow through ion selective membranes to electrodes of opposite charge. In EDR systems, the polarity of the electrodes is reversed periodically. Ion-transfer (perm-selective) anion and cation membranes separate the ions in the feed water. These systems are used primarily in waters with low total dissolved solids (TDS).

Forward osmosis (FO) is a relatively new commercial desalting process in which a salt concentration gradient (osmotic pressure) is the driving force through a synthetic membrane. The feed (such as seawater) is on one side of the semi permeable membrane and a higher osmotic pressure "draw" solution is on the other side. Without applying any external pressure, the water from the feed solution will naturally migrate through the membrane to the draw solution. The diluted solution is then processed to separate the product from the reusable draw solution.

Membrane distillation (MD) is a water desalination membrane process currently in limited commercial use. MD is a hybrid process of RO and distillation in which a hydrophobic synthetic membrane is used to permit the flow of water vapor through the membrane pores, but not the solution itself. The driving force for MD is the difference in vapor pressure of the liquid across the membrane.

Importance of Water Resources

Residential, commercial and industrial users often compete with agricultural irrigation, hydropower production and navigation for water resources.

Residential, Commercial and Industrial use

Residential water use includes drinking, cleaning, personal hygiene, lawn care and car washing. Americans get water from public water systems and from private supplies such as wells. In the commercial and industrial sectors, most water is used for processing products, with cooling coming in second. Water is also used for laundry, sanitation and landscaping. In 2000, industrial users consumed 5 percent of the available water in the United States.

Hydropower

Hydroelectric facilities use the power of flowing water to turn turbines that produce electricity. According to the Department of Energy, the U.S. produces 95,000 megawatts of hydropower per year -- enough to power 28 million households or replace 500 million barrels of oil. Hydropower has come under scrutiny from environmentalists, but new technologies promise to increase the efficiency of power generation while simultaneously decreasing the impact of hydroelectricity on the environment.

Irrigation

Only 15 percent of cropland in the United States is irrigated, but that still totals about 55 million acres, including land in highly productive areas such as California. Water for irrigation comes from either groundwater or surface water, raising concerns that heavy use could deplete water supplies in a region to the extent that nonagricultural users are negatively affected. Irrigation has also been linked to increased soil salinity and contamination of groundwater with fertilizers and chemicals through runoff.

Navigation

Navigable waterways are defined as watercourses that have been or may be used for transport of interstate or foreign commerce. Agricultural and commercial goods are moved on water on a large scale in the United States, making navigation an important economic concern. Federal regulations control construction, excavation and disposal in and around navigable waters. Navigation interests may come into direct conflict with other interests, including hydropower and wildlife conservation.

Irrigation

Irrigation is a broad term referring to any means of delivering water to growing plants. It can take a number of different forms, from irrigation ditches to drip irrigation and more. It also applies to maintaining landscaping features, such as turf/grass, trees, shrubs, and flowers.

Irrigation has essentially been used since humans first began cultivating plants. Any sort of cultivated plant requires water in order to grow and thrive. In many instances, rainfall is not sufficient to achieve this goal. In other instances, an area may experience short or prolonged periods of drought. Irrigation works to get around these problems.

There are quite a few forms of irrigation. The simplest is hand watering, which is simply pouring water over a plant and the soil surrounding it using a bucket or watering can. However, there are better ways of achieving this. Furrow or flood irrigation is the term used to describe irrigating crops using a series of furrows or ditches filled with water in the growing area to deliver water.

Drip irrigation relies on the use of drip emitters, or plastic pipes with tiny holes in them (drip hoses) designed specifically to "sweat" water out of them. This delivers a much more controlled amount of water to the plant, and reduces evaporation and transpiration.

Spray irrigation is yet another option, although this requires the use of machinery and is more expensive to implement. Spray irrigation relies on a series of hoses and sprayers or sprinklers – you're probably familiar with this from automatic lawn-watering systems.

A derivative of this system, called gentle spray irrigation, is used in areas where wind, a lack of humidity, or high heat levels may increase water waste during irrigation. Gently spray irrigation relies on gentle force and the use of hanging pipes above the growing area to deliver vital water.

Importance of Irrigation

1. Irrigation maintains moisture in the soil. Moisture is necessary for the germination of seeds. Seeds do not grow in dry soil. That is why irrigation is done before tilling.

2. Irrigation is essential for the growth of the roots of the crop plants. Roots of the plants do not grow well in dry soil.

3. Irrigation is necessary for the absorption of mineral nutrients by the plants from the soil. Thus, irrigation is essential for the general growth of the plants.

4. Water supplies two essential elements hydrogen and oxygen to the crop.

References

- Water-resource, science: britannica.com, Retrieved 24 March 2018
- Factsheet-what-is-surface-water: denr.nt.gov.au, Retrieved 16 May 2018
- Different-sources-water-7624072: sciencing.com, Retrieved 15 July 2018
- Water-Desalination-Processes: amtaorg.com, Retrieved 11 June 2018
- Uses-natural-water-resources-79287: homeguides.sfgate.com, Retrieved 31 March 2018

Chapter 2

Types of Irrigation

There are many methods of irrigation, such as surface irrigation, sprinkler irrigation, subirrigation and micro-irrigation. The topics elucidated in this chapter cover some of these important types of irrigation such as surface, drip, sprinkler, center pivot irrigation, etc.

Surface Irrigation

Surface irrigation has evolved into an extensive array of configurations, which can be broadly classified as:

(1) Basin irrigation.

(2) Border irrigation.

(3) Furrow irrigation.

(4) Uncontrolled flooding.

There are two features that distinguish a surface irrigation system:

(a) The flow has a free surface responding to the gravitational gradient.

(b) The on-field means of conveyance and distribution is the field surface itself.

A surface irrigation event is composed of four phases as illustrated graphically in figure below. When water is applied to the field, it 'advances' across the surface until the water extends over the entire area. It may or may not directly wet the entire surface, but all of the flow paths have been completed. Then the irrigation water either runs off the field or begins to pond on its surface. The interval between the end of the advance and when the inflow is cut off is called the wetting or ponding phase. The volume of water on the surface begins to decline after the water is no longer being applied. It either drains from the surface (runoff) or infiltrates into the soil. For the purposes of describing the hydraulics of the surface flows, the drainage period is segregated into the depletion phase (vertical recession) and the recession phase (horizontal recession). Depletion is the interval between cut off and the appearance of the first bare soil under the water. Recession begins at that point and continues until the surface is drained.

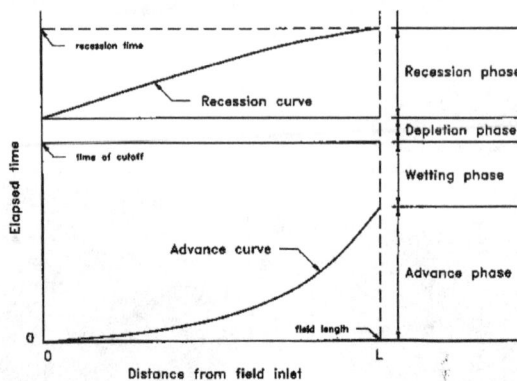

Time-space trajectory of water during a surface irrigation showing its advance, wetting, and depletion and recession phases

The time and space references shown in figure above are relatively standard. Time is cumulative since the beginning of the irrigation, distance is referenced to the point water enters the field. The advance and recession curves are therefore trajectories of the leading and receding edges of the surface flows and the period defined between the two curves at any distance is the time water is on the surface and therefore also the time water is infiltrating into the soil.

It is useful to note here that in observing surface irrigation one may not always observe a ponding, depletion or recession phase. In basins, for example, the post-cut off period may only involve a depletion phase as the water infiltrates vertically over the entire field. Likewise, in the irrigation of paddy rice, irrigation very often adds to the ponded water in the basin so there is neither advance nor recession - only wetting or ponding phase and part of the depletion phase. In furrow systems, the volume of water in the furrow is very often a small part of the total supply for the field and it drains rapidly. For practical purposes, there may not be a depletion phase and recession can be ignored. Thus, surface irrigation may appear in several configurations and operate under several regimes.

The surface irrigation system is one component of a much larger network of facilities diverting and delivering water to farmlands. Figure above illustrates the 'irrigation system' and some of its features. It may be divided into the following four component systems:

- Water supply.

- Water conveyance or delivery.

- Water use.

- Drainage.

For the complete system to work well, each must work conjunctively toward the common goal of promoting maximum on-farm production. Historically, the elements of an irrigation system have not functioned well as a system and the result has too often been very low project irrigation efficiencies.

The focus of surface irrigation engineering is at the water use level, the individual irrigated field. For design and evaluation purposes, these guidelines will note elements of the conveyance and distribution system, especially those near the field such as flow measurement and control, but will leave detailed treatment to other technical sources.

Evolution of the Practice

Although surface irrigation is thousands of years old, the most significant advances have been made within the last decade. In the developed and industrialized countries, land holdings have become as much as 10-20 times as large, and the number of farm families has dropped sharply. Very large mechanized farming equipment has replaced animal-powered planting, cultivating and harvesting operations. The precision of preparing the field for planting has improved by an order of magnitude with the advent of the laser-controlled land grading equipment. Similarly, the irrigation works themselves are better constructed because of the application of high technology equipment.

The changes in the lesser-developed and developing countries are less dramatic. In the lesser-developed countries, trends toward land consolidation, mechanization, and more elaborate system design and operation are much less apparent. Most of these farmers own and operate farms of 1-10 hectares, irrigate with 20-40 liters per second and rely on either small mechanized equipment or animal-powered farming implements.

Probably the most interesting evolution in surface irrigation so far as this guide is concerned is the development and application of microcomputers and programmable calculators to the design and operation of surface irrigation systems. In the late 1970s, a high-speed microcomputer technology began to emerge that could solve the basic equations describing the overland flow of water quickly and inexpensively. At about the same time, researchers like Strelkoff and Katapodes made major contributions with efficient and accurate numerical solutions to these equations. Today in the graduate and undergraduate study of surface irrigation engineering, microcomputer and programmable calculator utilization is, or should be, common practice.

Microcomputers and programmable calculators provide several features for today's irrigation engineers and technicians. They allow a much more comprehensive treatment of the vital hydraulic processes occurring both on the surface and beneath it. One can find optimal designs and management practices for a multitude of conditions because designs historically requiring days of effort are now made in seconds. The effectiveness of existing practices or proposed ones can be predicted, even to the extent that control systems operating, sensing and adjusting on a real-time basis are possible.

Surface Irrigation Methods

The classification of surface methods is perhaps somewhat arbitrary in technical literature. This has been compounded by the fact that a single method is often referred to with different names. In this guide, surface methods are classified by the slope, the size and shape of the field, the end conditions, and how water flows into and over the field.

Each surface system has unique advantages and disadvantages depending on such factors as were listed earlier like:

- Initial cost.

- Size and shape of fields.

- Soil characteristics.

- Nature and availability of the water supply.

- Climate.

- Cropping patterns.

- Social preferences and structures.

- Historical experiences.

- Influences external to the surface irrigation system.

Basin Irrigation

Basin irrigation is the most common form of surface irrigation, particularly in regions with layouts of small fields. If a field is level in all directions, is encompassed by a dyke to prevent runoff, and provides an undirected flow of water onto the field, it is herein called a basin. A basin is typically square in shape but exists in all sorts of irregular and rectangular configurations. It may be furrowed or corrugated, have raised beds for the benefit of certain crops, but as long as the inflow is undirected and uncontrolled into these field modifications, it remains a basin.

There are few crops and soils not amenable to basin irrigation, but it is generally favored by moderate to slow intake soils, deep-rooted and closely spaced crops. Crops, which are sensitive to flooding, and soils which form a hard crust following an irrigation can be basin irrigated by adding furrowing or using raised bed planting. Reclamation of salt-affected soils is easily accomplished with basin irrigation and provision for drainage of surface runoff is unnecessary. Of course it is always possible to encounter a heavy rainfall or mistake the cut-off time thereby having too much water in the basin. Consequently, some means of emergency surface drainage is good design practice. Basins can be served with less command area and field watercourses than can border and furrow systems because their level nature allows water applications from anywhere along the basin perimeter. Automation is easily applied.

Basin irrigation has a number of limitations, two of which, already mentioned, are associated with soil crusting and crops that cannot accommodate inundation. Precision land leveling is very important to achieving high uniformities and efficiencies. Many basins are so small that precision equipment cannot work effectively. The perimeter dykes need to be well maintained to eliminate breaching and waste, and must be higher for basins than other surface irrigation methods. To reach maximum levels of efficiency, the flow per unit width must be as high as possible without causing erosion of the soil. When an irrigation project has been designed for either small basins or furrows and borders, the capacity of control and outlet structures may not be large enough to improve basins.

Border Irrigation

Border irrigation can be viewed as an extension of basin irrigation to sloping, long rectangular or contoured field shapes, with free draining conditions at the lower end. Water is applied to individual borders from small hand-dug checks from the field head ditch. When the water is shut off, it recedes from the upper end to the lower end. Sloping borders are suitable for nearly any crop except those that require prolonged ponding. Soils can be efficiently irrigated which have moderately low to moderately high intake rates but, as with basins, should not form dense crusts unless provisions are made to furrow or construct raised borders for the crops. The stream size per unit width must be large, particularly following a major tillage operation, although not so large for basins owing to the effects of slope. The precision of the field topography is also critical, but the extended lengths permit better leveling through the use of farm machinery.

Furrow Irrigation

Furrow irrigation avoids flooding the entire field surface by channeling the flow along the primary direction of the field using 'furrows,' 'creases,' or 'corrugations'. Water infiltrates through the wetted perimeter and spreads vertically and horizontally to refill the soil reservoir. Furrows are often employed in basins and borders to reduce the effects of topographical variation and crusting. The distinctive feature of furrow irrigation is that the flow into each furrow is independently set and controlled as opposed to furrowed borders and basins where the flow is set and controlled on a border by border or basin by basin basis.

Furrows provide better on-farm water management flexibility under many surface irrigation conditions. The discharge per unit width of the field is substantially reduced and topographical variations can be more severe. A smaller wetted area reduces evaporation losses. Furrows provide the irrigator more opportunity to manage irrigations toward higher efficiencies as field conditions change for each irrigation throughout a season. This is not to say, however, that furrow irrigation enjoys higher application efficiencies than borders and basins.

There are several disadvantages with furrow irrigation. These may include:

- An accumulation of salinity between furrows.

- An increased level of tailwater losses.

- The difficulty of moving farm equipment across the furrows.

- The added expense and time to make extra tillage practice (furrow construction).

- An increase in the erosive potential of the flow.

- A higher commitment of labor to operate efficiently.

- Generally furrow systems are more difficult to automate, particularly with regard to regulating an equal discharge in each furrow.

(a) Graded furrow irrigation system

(b) Contour furrows

Uncontrolled Flooding

There are many cases where croplands are irrigated without regard to efficiency or uniformity. These are generally situations where the value of the crop is very small or the field is used for grazing or recreation purposes. Small land holdings are generally not subject to the array of surface irrigation practices of the large commercial farming systems. Also in this category are the surface irrigation systems like check-basins which irrigate individual trees in an orchard, for example. The evaluation methods can be applied if desired, but the design techniques are not generally applicable nor need they be since the irrigation practices tend to be minimally managed.

Requirements for Optimal Performance

There is substantial field evidence that surface irrigation systems can apply water to croplands uniformly and efficiently, but it is the general observation that most such systems operate well below their potential. A very large number of causes of poor surface irrigation performance have been outlined in the technical literature. They range from inadequate design and management at the farm level to inadequate operation of the upstream water supply facilities. However, in looking for a root cause, one most often retreats to the fact that infiltration changes a great deal from irrigation to irrigation, from soil to soil, and is neither predictable nor effectively manageable. The infiltration rates are an unknown variable in irrigation practice.

In those cases where high levels of uniformity and efficiency are being achieved, irrigators utilize one or more of the following practices:

- Precise and careful field preparation.

- Irrigation scheduling.

- Regulation of inflow discharges.

- Tail water runoff restrictions, reduction, or reuse.

Inlet Discharge Control

Surface irrigation systems have two principal sources of inefficiency, deep percolation and surface runoff or tail water. The remedies are competitive. To minimize deep percolation the advance phase should be completed as quickly as possible so that the intake opportunity time over the field will be uniform and then cut the inflow off when enough water has been added to refill the root zone. This can be accomplished with a high, but non-erosive, discharge onto the field. However, this practice increases the tail water problem because the flow at the downstream end must be maintained until a sufficient depth has infiltrated. The higher inflow reaches the end of the field sooner but it increases both the duration and the magnitude of the runoff.

There are three options available to solve this problem, at least partially:

- Dyke the downstream end to prevent runoff as in basin irrigation.

- Reduce the inflow discharge to a rate more closely approximating the cumulative infiltration along the field following the advance phase, a practice termed 'cutback'.

- Select a discharge, which minimizes the sum of deep percolation and tail water losses, i.e., optimize the field inflow regime.

In this configuration, the head ditch is divided into a series of level bays, which are differentiated by a small change in elevation. Water levels are regulated in two bays simultaneously so that the lower bay has sufficient head to produce an advance phase flow in the furrows while in the upper bay the head is only sufficient to produce the cutback flow. Thus, the system operates by moving the check-dam from bay to bay along the upper end of the field.

Two very recent additions to the efforts to control surface irrigation systems more effectively are the 'Surge Flow' system developed at Utah State University, USA and the 'Cablegation' system developed at the US Department of Agriculture's Snake River Water Conservation Research Center in Kimberly, Idaho, USA.

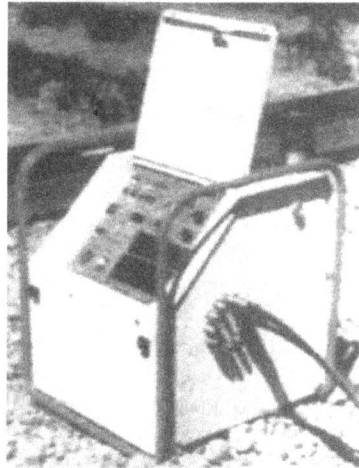

Wastewater Recovery and Reuse

The tail water deep percolation trade-off can also be solved by collecting and recycling the runoff to improve surface irrigation performance. Reuse systems have not been widely employed historically because water and energy have been inexpensive. Even today it is often more economical to regulate the inflow rather than to collect and pump the runoff back to the head of the field or to another field, tail water reuse systems are more cost-effective when the water can be added to the flow serving lower fields and thereby saving the cost of pumping.

Surface Irrigation Structures

Surface irrigation systems are supported by a number of on- and off-farm structures which control and manage the flow and its energy. In order to facilitate efficient surface irrigation, these structures should be easily and cheaply constructed as well as easy to manage and maintain. Each should be standardized for mass production and fabrication in the field by farmers and technicians.

It is not the intent of this guide to be comprehensive with regard to the selection and design of these structures since other sources are available, but it is worthwhile to note some of these structures by way of presenting a larger view of surface irrigation.

The structural elements of a surface system perform several important functions, which include:

- Turning the flow to a field on and off.

- Conveying and distributing the flow among fields.

- Water measurement, sediment and debris removal, and water level stabilization.

- Distribution of water onto the field.

Diversion Structures

Most surface irrigation systems derive their water supplies from canal systems operated by public or semi-public irrigation departments, districts, or companies. Some irrigation water is supplied in piped delivery systems and some directly pumped from groundwater. Diversion structures perform several tasks including:

- On-off water control, which allows the supply agency to allocate its supply, and protects the fields below the diversion from untimely flooding.

- Regulation and stabilization of the discharge to the requirements of field channels and watercourse distribution systems.

- Measurement of flow at the turnout in order to establish and protect water entitlements.

- Protection of downstream structures by controlling sediments and debris as well as dissipating excess kinetic energy in the flow. A typical turnout structure is shown in figure below.

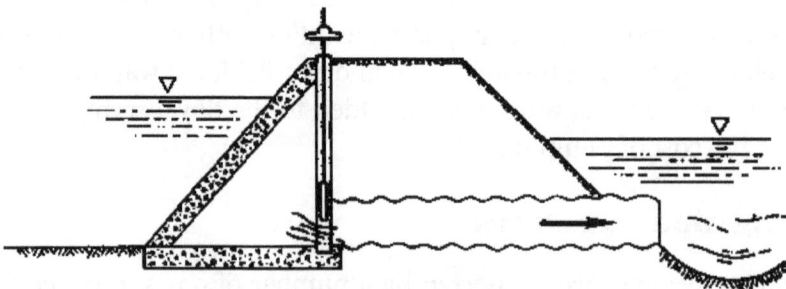

Typical turnout from a canal or lateral

Conveyance, Distribution and Management Structures

Conveying water to the field requires similar structures to those found in major canal networks. The conveyance itself can be an earthen ditch or lateral, a buried pipe, or a lined ditch. Lined sections can be elevated or constructed at surface level. Pipe materials are usually plastic, steel, concrete, clay, or asbestos cement, or they may be as simple as a wooden or bamboo construction. Lining materials include slip-form cast-in-place, or prefabricated concrete, shotcrete or gunite, asphalt, surface and buried plastic or rubber membranes, and compacted earth.

The management of water in the field channels involves flow measurement, sediment and debris removal, divisions, checks, drop-energy dissipators, and water level regulators. Some of the more common flow control structures for open channels are shown in figure below. Associated with these are various flow measuring devices like weirs, flumes, and orifices. The designs of these structures have been standardized since they are small in size and capacity. Designs for flow measurement and drop-energy dissipator structures need more attention and construction must be more precise since their hydraulic responses are quite sensitive to their dimensions.

The figure below shows on-farm water management structures

A simple drop structure   A typical check-divider

Field Distribution Systems

After the water reaches the field ready to be irrigated, it is distributed onto the field by a variety of means, both simple and elaborately constructed. Most fields have a head ditch or pipeline running along the upper side of the field from which the flow is distributed onto the field.

In a field irrigated from a head ditch, the spreading of water over the field depends somewhat on the method of surface irrigation. For borders and basins, open or piped cutlets are generally used. Furrow systems use outlets which can be directed to each furrow.

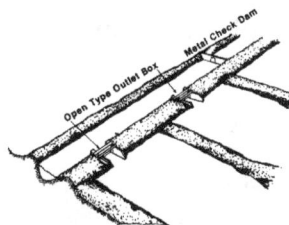

Head ditch outlets for borders and basins

Field distribution and spreading can also be through portable pipelines running along the surfaces or permanent pipelines running underground. Basins and borders usually receive water through buried pipes serving one or more gated risers within each basin or border. A typical riser outlet, known as an alfalfa valve, is shown in figure. The most common piped method of furrow irrigation uses plastic or aluminium gated pipe like that shown in figure. The gated pipe may be connected to the main water supply via a piped distribution network with a riser assembly like the one shown in figure, directly to a canal turnout, or through an open channel to a piped transition.

An alfalfa valve riser

Benefits

- Low cost of deployment, energy and maintenance.

- It favors the increase of the photosynthesis in the lower leaves, due to the reflection of the light in the water.

- The wind does not limit the irrigation.

- It promotes the fixation of atmospheric nitrogen, as a result of favoring the growth of blue-green algae.

Disadvantages

- Standing water can harm plants, mainly by reducing the respiration of the roots.

- Fairly dependent on soil slope.

- Erosions frequent in the grooves.

- Loss of water occurs due to percolation.

Micro-irrigation

Conventional irrigation systems, such as channel irrigation and wild flooding tend to waste water as large quantities are supplied to the field in one go, most of which just flows over the crop and runs away without being taken up by the plants.

Micro irrigation is an approach to irrigation that keeps the water demand to a minimum. It has been driven by commercial farmers in arid regions of the United States of America and Israel in farming areas where water is scarce.

Typically, these commercial irrigation systems consist of a surface or buried pipe distribution network using emitters supplying water directly to the soil at regular intervals along the pipework. They can be permanent or portable.

Kenya Practical Action staff demonstrate drip irrigation

Many parts of the world are now using micro irrigation technology. The systems used by large commercial companies are generally quite complex with an emphasis on reducing the amount of labor involved.

Small-scale farmers in developing countries have been reluctant to take up micro irrigation methods due to the initial investment required for the equipment. A number of organizations have looked at ways to simplify and reduce the cost of micro irrigation resulting in the approaches of drip irrigation and pipe irrigation. For these small-scale irrigation systems not only should the technical aspects of the system be considered such as:

- Access to reliable water sources
- A secure and well-fenced garden
- Basic gardening skills
- The crops grown

The social aspects should also be of concern to ensure the irrigation system will be of benefit. The social and economical factors will include:

- capital and financial management; credit facilities

- the availability of external services

- maintenance

- market opportunities for the produce

- willingness to show other farmers the technology

Drip Irrigation

Drip irrigation uses low-cost plastic pipes laid on the ground to irrigate vegetables, field crops and orchards. This technology was developed in the 1960s for commercial use. Circa 1990 a US firm called Chapin Watermatics developed a low-cost system called bucket kits, which use standard plastic buckets and lengths of hose that could be cut to the appropriate lengths.

Small holes in the hose allow water to drip out and keep the base of the plant wet without wasting any water. The kits are low-cost, easy to assemble and manage. They do not need high quality water, providing the water is filtered. A 20 liter bucket with 30 meters (100 feet) of hose or drip tape connected to the bottom. The bucket is placed at least 1 meter (3 feet) above the ground so that gravity provides sufficient water pressure to ensure even watering for the entire crop. Water is poured into the bucket twice daily and passes through a filter, fills the drip tape and is evenly distributed to 100 watering points. The multi-chambered plastic drip tape is engineered to dispense water through openings spaced at 30cm (12 inches).

Two bucket kits costing around $20 will produce enough vegetables for a family of seven and can last over five years. The system is most suited to kitchen gardens. As well as the bucket, you will need several strong poles, tools, manure, water and vegetable seedlings. The poles are used to make a support structure for the bucket. The stand should hold the bucket about 1 meter above the ground. The main stages of setting up the systems are:

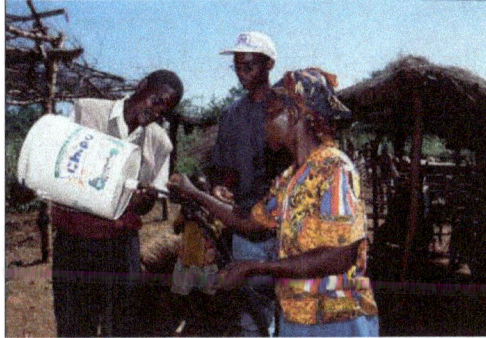

Practical Action farmers project staff examine the drip irrigation kit.

- A hole should be cut carefully into the base of the bucket.

- The hole is fitted with the filter plug and tubing and then flushed out to ensure that the system is clean.

- The drip lines are then connected and the system is flushed out again before the ends of the drip lines are closed off.

The whole procedure of setting up the system will only take about one hour, including the construction of the bucket support. When planting a seedling is planted at each wet spot so that all the moisture is absorbed directly by the plant roots. Moving the kit from plot to plot tends not to be very practical and damages the equipment. It is better to add extra buckets and lines when necessary or when funds are available to invest in additional equipment.

The advantages are:

- The effort to water the plants is greatly reduced

- The time taken to fill the containers is significantly less than manually watering the planted area.

- The growth of weeds is reduced, as water does not reach unwanted plants.

Some initial training is required to show how to get the best results from the system and careful attention should be taken to a number of common problems that occur with the system comprising of:

- Clogging of the trip tapes especially if water is not well filtered. The emitters can be cleaned by blowing dirt out and then flushing out the particles with clean water.

- Leakage at connections: this can be checked and corrected.

- Weeding must avoid puncturing the tape.

- Vermin can damage pipes in their search for water.

- Theft of equipment.

The bucket kit is the smallest type of drip irrigation available. Although productivity is increased providing greater food security and improved nutrition, the amount of water and labour saved is small. Often there is not sufficient surplus produce for users to sell and provide financial benefits.

Bigger containers can be used to suit larger market gardens. Large customized drum kits irrigating a high-value crop offer greatest financial impact.

In East Africa drip irrigation has been promoted by Practical Action East Africa and the Arid Land Information Network (ALIN) who sell drip irrigation kits similar to the ones used in India and Zimbabwe.

Water is poured into the pipe which then irrigates the plants at the roots

Pipe Irrigation

The use of buried clay pitchers is an ancient technique of subsurface irrigation. The use of clay pipes had been initially tried in Russia and Mexico. ITDG Southern Africa along with others in Zimbabwe developed a low cost variation of this irrigation method in which clay pipes were buried beneath vegetable beds of the 450 women members of garden groups.

One end of the pipe is blocked and the other is tilted out of the soil to allow filling. The pipes were laid end to end below the soil surface. When the pipe is filled, water will gradually escape from the cracks between the pipe sections and through the pores in the clay to provide a continuous supply of water to the vegetables.

There were difficulties in obtaining the correct dimensions. Research at the Lowveld Research Station, Zimbabwe found that the optimum pipe size was 75mm inside diameter and 300mm long.

The pipes are produced using a mould made from a plastic drainpipe with handles attached. The clay is prepared by rolling it out to a suitable size and thickness before it is put into the mould. Once it has been fitted to the mould the pipe can be removed and stood on end to dry naturally. The pipes are fired in an open fire.

Although the pipes are still laborious to fill by bucket, the use of clay pipes reduced the work for the women because the vegetable gardens needed to be watered only once a week instead of 3-4 times. Watering was said to be reduced by 50% and vegetable production increased.

The main benefit of the clay pipe approach is that the pipes were manufactured by the women, who ran the vegetable garden, without any external input. The disadvantages are that it is a long and laborious task to produce, that the installation of clay pipes has to be carried out every season, and the pipes are easily broken during garden operations.

Need of Micro Irrigation

- To make agriculture productive,

- Environmentally sensitive and capable of preserving the social fabric of rural communities

- Help produce more from the available land, water and labor resources without either ecological or social harmony,

- Generate higher farm income

- On-farm and off-farm employment.

Drip Irrigation

Drip irrigation is sometimes called trickle irrigation and involves dripping water onto the soil at very low rates (2-20 liters/hour) from a system of small diameter plastic pipes fitted with outlets called emitters or drippers. Water is applied close to plants so that only part of the soil in which the roots grow is wetted, unlike surface and sprinkler irrigation, which involves wetting the whole soil profile. With drip irrigation water, applications are more frequent (usually every 1-3 days) than with other methods and this provides a very favorable high moisture level in the soil in which plants can flourish.

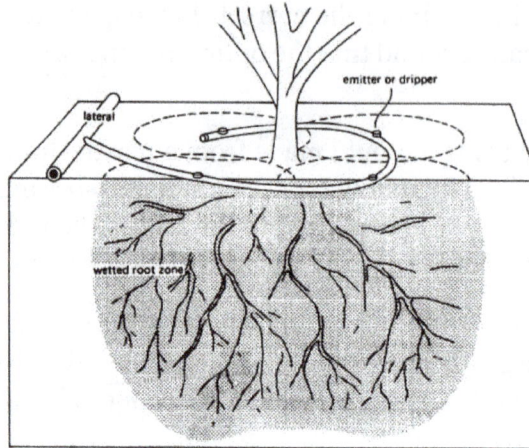

Drip irrigation system delivers water to the crop using a network of mainlines, sub-mains and lateral lines with emission points spaced along their lengths. Each dripper/emitter, orifice supplies a measured, precisely controlled uniform application of water, nutrients, and other required growth substances directly into the root zone of the plant.

Water and nutrients enter the soil from the emitters, moving into the root zone of the plants through the combined forces of gravity and capillary. In this way, the plant's withdrawal of moisture and nutrients are replenished almost immediately, ensuring that the plant never suffers from water stress, thus enhancing quality, its ability to achieve optimum growth and high yield.

Drip System Layout

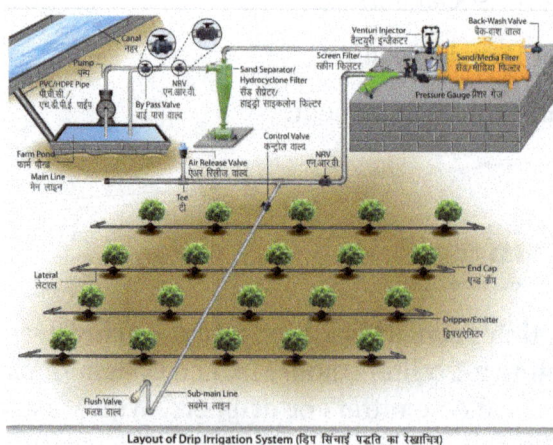

Layout of Drip Irrigation System (ड्रिप सिंचाई पद्धति का रेखाचित्र)

Major Components of Drip Irrigation System			
1	Pump station	2	By-pass assembly
3	Control valves	4	Filtration system

5	Fertilizer tank /Venturi	6	Pressure gauge
7	Mains / Sub-mains	8	Laterals
9	Emitting devices	10	Micro tubes

- Pump station takes water from the source and provides the right pressure for delivery into the pipe system.

- Control valves control the discharge and pressure in the entire system.

- Filtration system cleans the water. Common types of filter include screen filters and graded sand filters, which remove fine material suspended in the water.

- Fertilizer tank/venturi slowly add a measured dose of fertilizer into the water during irrigation. This is one of the major advantages of drip irrigation over other methods.

- Mainlines, submains and laterals supply water from the control head into the fields. They are usually made from PVC or polyethylene hose and should be buried below ground because they easily degrade when exposed to direct solar radiation. Lateral pipes are usually 13-32 mm diameter.

- Emitters or drippers are devices used to control the discharge of water from the lateral to the plants. They are usually spaced more than 1 meter apart with one or more emitters used for a single plant such as a tree. For row crops more closely spaced emitters may be used to wet a strip of soil. Many different emitter designs have been produced in recent years. The basis of design is to produce an emitter, which will provide a specified constant discharge which does not vary much with pressure changes, and does not block easily.

Wetting Pattern in Drip Irrigation

Unlike surface and sprinkler irrigation, drip irrigation only wets part of the soil root zone. This may be as, low as 30% of the volume of soil wetted by the other methods. The wetting patterns which develop from dripping water onto the soil depend on discharge and soil type. Figure below shows the effect of changes in discharge on two different soil types, namely sand and clay.

Although only part of the root zone is wetted it is still important to meet the full water needs of the crop. It is sometimes thought that drip irrigation saves water by reducing the amount used by the crop. This is not true. Crop water use is not changed by the method of applying water. Crops just require the right amount for good growth.

The water savings that can be made using drip irrigation are the reductions in deep percolation, in surface runoff and in evaporation from the soil. These savings, it must be remembered, depend as much on the user of the equipment as on the equipment itself.

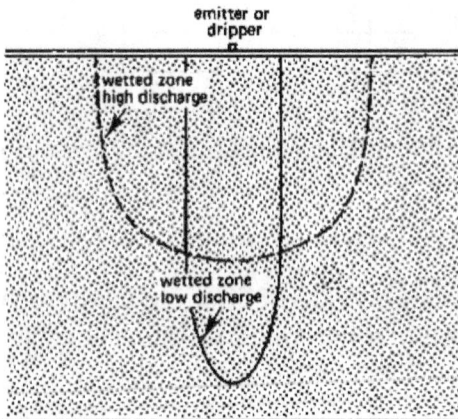

Wetting pattern in Sandy Soils   Wetting pattern in Clay Soils

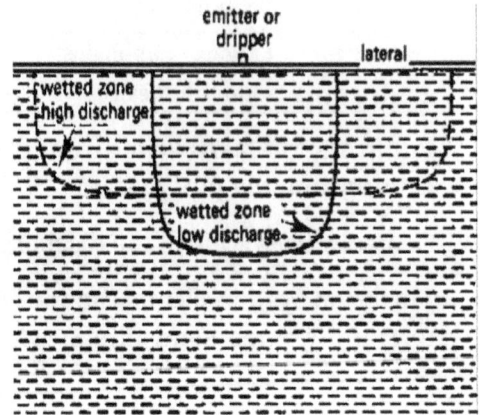

Drip irrigation is not a substitute for other proven methods of irrigation. It is just another way of applying water. It is best suited to areas where water quality is marginal, land is steeply sloping or undulating and of poor quality, where water or labor are expensive, or where high value crops require frequent water applications.

Crops Suitable for Drip Irrigation System

1	Orchard Crops	Grapes, Banana, Pomegranate, Orange, Citrus, Mango, Lemon, Custard Apple, Sapota, Guava, Pineapple, Coconut, Cashewnut, Papaya, Aonla, Litchi, Watermelon, Muskmelon etc.
2	Vegetables	Tomato, Chilly, Capsicum, Cabbage, Cauliflower, Onion, Okra, Brinjal, Bitter Gourd, Ridge Gourd, Cucumber, Peas, Spinach, Pumpkin etc.
3	Cash Crops	Sugarcane, Cotton. Arecanut, Strawberry etc.
4	Flowers	Rose, Carnation, Gerbera, Anthurium, Orchids, Jasmine, Dahilia, Marigold etc.
5	Plantation	Tea, Rubber, Coffee, Coconut etc.
6	Spices	Turmeric, Cloves, Mint etc,
7	Oil Seed	Sunflower, Oil palm, Groundnut etc.
8	Forest Crops	Teakwood, Bamboo etc.

Response of Different Crops to Drip Irrigation System

Crops	Water saving (%)	Increase in yield (%)
Banana	45	52
Cauliflower	68	70
Chilly	68	28
Cucumber	56	48
Grapes	48	23
Ground nut	40	152

Pomegranate	45	45
Sugarcane	50	99
Sweet lime	61	50
Tomato	42	60
Watermelon	66	19

Benefits of Drip Irrigation

- Increase in yield up to 230 %.

- Saves water up to 70% compare to flood irrigation. More land can be irrigated with the water thus saved.

- Crop grows consistently, healthier and matures fast.

- Early maturity results in higher and faster returns on investment.

- Fertilizer use efficiency increases by 30%.

- Cost of fertilizers, inter-culturing and labor use gets reduced.

- Fertilizer and Chemical Treatment can be given through Micro Irrigation System itself.

- Undulating terrains, Saline, Water logged, Sandy & Hilly lands can also be brought under productive cultivation.

Water Conservation Through Drip

Water is conserved in the following ways:

- Drip irrigation application uniformity is very high, usually over 90%.

- Unlike sprinklers, drip irrigation applies water directly to the soil, eliminating water loss from wind.

- Application rates are low so water may be spoon fed to the crop or plant root zone in the exact amounts required (even on a daily or hourly basis). In contrast, other methods entail higher water application quantities and less frequency. If young plants need water frequently, much of the water applied is often wasted to deep percolation or runoff.

- Low application rates are less likely to run off from heavier soils or sloping terrain.

- Drip irrigation does not water non-targeted areas such as furrows and roads in agriculture, between beds, blocks or benches in greenhouses, or hardscape, buildings or roads in landscape.

- Drip irrigation easily adapts to odd-shaped planting areas which are difficult to address with sprinklers or gravity irrigation.

- Drip irrigation is capable of germinating seeds and setting transplants which eliminates the need for "sprinklering up" and eliminates the resulting waste in the early stages of crop growth.

Drip irrigation is today's need because Water - nature's gift to mankind is not unlimited and free forever. World water resources are fast diminishing.

A Case Study of the Earlier Approaches to Drip Irrigation

Dating back 200 years, tribes have used bamboo drip irrigation as a means of bringing water to seasonal crops. This timeless and traditional technology uses locally available material while harnessing the forces of gravity. An assortment of holed bamboo shoots zig-zag downhill, diverting the natural flow of streams and springs across terraced cropland. The advantages of using bamboo are two-fold: it prevents leakage, increasing crop yield with less water, and makes use of natural, local, and inexpensive material.

The Jaintia, Khasi, and Garo hills of Meghalaya are largely made up of steep slopes and generally rocky terrain where the soil has low water retention capacity and where the use of groundwater channels is impossible. During the dry seasons, rainfed crops such as paddy, betal leaf, and black peppers can be irrigated by bamboo drip irrigation. Within the Jainta hills, the small village of Nongbah relies on terrace agriculture for paddy cultivation. There are no restrictions for individuals tapping into water flows from perennial streams, natural springs, or collection ponds. This enables farmers, nearly 97% of the population, to cultivate paddy, betal leaf, and black peppers in seasonal rotations. Meanwhile, drinking needs are met by perennial springs during the dry months, from October to March.

Only during the winter is irrigation required, and the bamboo system is used for crops that need relatively less water.

The few materials needed are a small dao (a type of local axe), bamboo strands of various sizes, forked branches, smaller bamboo shoots used for the channel diversions, and two willing laborers. A. Singh, in his book Bamboo Drip Irrigation Systems, investigated its use in Meghalaya and says that two workers can construct a system covering one hectare of land in 15 days. About four or five stages of irrigation zig-zag from the water source to the last point of application. Along the way, 18-20 liters of water will eventually disseminate at a rate of 20-80 drops per minute.

To get started, locate an available water source. Next, select a sloped area of land (at least 30 meters in variation). Then slice the bamboo shoots and forked branches, placing the wider shoots in the first channel and the smaller pipes for the last section (plan for 5 stages). Puncture a series of holes in the shoots, spacing them equally. Ground clearance should progressively descend so that the water may be dropped near the roots of plants in the last section (10-15 cm above ground). To reinforce the structure, tie the pipes and forked branches together using fiber-rich twine as rope. At points of diversion, smaller bamboo shoots may be used to redirect water.

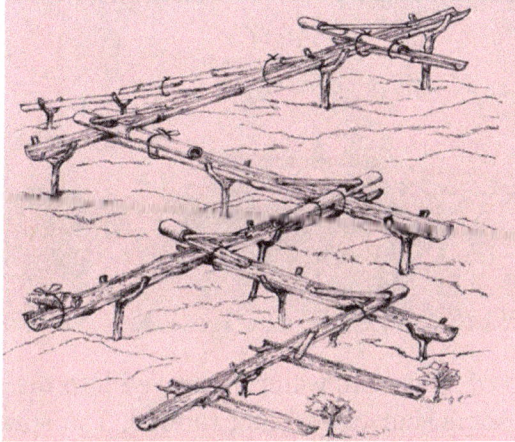

Bamboo drip irrigation system

Materials used during installation last around three years, while maintenance is limited to cleaning and reinforcement after seasonal monsoons. Cost is also limited to labor, which can be carried out by farmers themselves. This traditional water structure is indeed contextual to location. Such variables include: replenishable supplies of bamboo, upland sources of water, and the presence of traditional terrace agriculture.

Adapting to drier growing seasons, farmers are advised to match irrigation decisions with crop selection. It is suggested that during the June to September season, rice cultivation can be stopped and farmers can opt for crops such as pulses, sunflower, sesame and maize. In areas affected severely by drought, farmers may go for pearl millet, minor millet and forage crops. However, in places where there is water stagnation, they can continue to grow rice.

In the Meghalaya hills there are an estimated 3,108 square kilometers of bamboo forests. In 1990, it was estimated that the total yield of bamboo in the state was 2.09 tons/hectare/year. The 38 different species of bamboo in the region are typically harvested at the community level and loosely managed by the Autonomous District Councils (ADC). Bamboo supplies have recently come under threat from a surge in rodent populations, gregarious flowering, disease, and large-scale extraction. Nonetheless, large-scale conservation and protection plans are underway in Arunachal Pradesh, Assam, Manipur, Meghalaya, Nagaland, Mizoram, Tripura, and Sikkim, which together contain over 50% of India's bamboo supply (within 226,000 square kilometers of land.)

The Jaintia, Khasi, and Garo hill tribes have long entrusted the use of bamboo drip irrigation as a means to fulfilling domestic, agricultural, and customary needs. Its function remains unspoiled so as the rains continue to fall and the bamboo continues to grow.

Sprinkler Irrigation

In the sprinkler method of irrigation, water is sprayed into the air and allowed to fall on the ground surface somewhat resembling rainfall. The spray is developed by the flow of water under pressure through small orifices or nozzles. The pressure is usually obtained by pumping. With careful selection of nozzle sizes, operating pressure and sprinkler spacing the amount of irrigation water required to refill the crop root zone can be applied nearly uniform at the rate to suit the infiltration rate of soil.

Crop Response to Sprinkler

The trials conducted in different parts of the country revealed water saving due to sprinkler system varies from 16 to 70 % over the traditional method with yield increase from 3 to 57 % in different crops and agro climatic conditions.

Response of Different Crops to Sprinkler Irrigation

Crops	Water Saving, %	Yield increase, %
Bajra	56	19
Barley	56	16
Bhindi	28	23
Cabbage	40	3
Cauliflower	35	12
Chillies	33	24
Cotton	36	50
Cowpea	19	3
Fenugreek	39	35
Garlic	28	6
Gram	69	57
Groundnut	20	40

Jowar	55	34
Lucerne	16	27
Maize	41	36
Onion	33	23
Potato	46	4
Sunflower	33	20
Wheat	35	24

Suitable Slopes

Sprinkler irrigation is adaptable to any farmable slope, whether uniform or undulating. The lateral pipes supplying water to the sprinklers should always be laid out along the land contour whenever possible. This will minimize the pressure changes at the sprinklers and provide a uniform irrigation.

Suitable Soils

Sprinklers are best suited to sandy soils with high infiltration rates although they are adaptable to most soils. The average application rate from the sprinklers (in mm/hour) is always chosen to be less than the basic infiltration rate of the soil so that surface ponding and runoff can be avoided.

Sprinklers are not suitable for soils which easily form a crust. If sprinkler irrigation is the only method available, then light fine sprays should be used. The larger sprinklers producing larger water droplets are to be avoided.

Suitable Irrigation Water

A good clean supply of water, free of suspended sediments, is required to avoid problems of sprinkler nozzle blockage and spoiling the crop by coating it with sediment.

Sprinkler System Layout

A typical sprinkler irrigation system consists of the following components:

- Pump unit
- Mainline and sometimes submainlines
- Laterals
- Sprinklers

The pump unit is usually a centrifugal pump, which takes water from the source and provides adequate pressure for delivery into the pipe system.

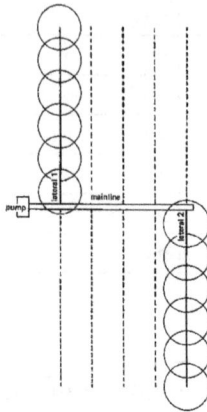

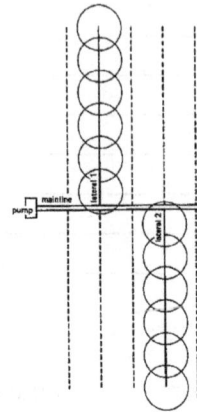

Hand-moved sprinkler system using two   Hand-moved sprinkler system using
laterals (Laterals 1 and 2 in position 1)   two laterals (Laterals 1 and 2 in position 2)

The mainline and submainlines are pipes which deliver water from the pump to the laterals. In some cases these pipelines are permanent and are laid on the soil surface or buried below ground. In other cases they are temporary, and can be moved from field to field. The main pipe materials used include asbestos cement, plastic or aluminium alloy.

The laterals deliver water from the mainlines or submainlines to the sprinklers. They can be permanent but more often they are portable and made of aluminium alloy or plastic so that they can be moved easily.

The most common type of sprinkler system consists of a system of lightweight aluminium or plastic pipes which are moved by hand. The rotary sprinklers are usually spaced 9-24 m apart along the lateral which is normally 5-12.5 cm in diameter. This is so it can be carried easily. The lateral pipe is located in the field until the irrigation is complete. The pump is then switched off and the lateral is disconnected from the mainline and moved to the next location. It is re-assembled and connected to the mainline and the irrigation begins again. The lateral can be moved one to four times a day. It is gradually moved around the field until the whole field is irrigated. This is the simplest of all systems. Some use more than one lateral to irrigate larger areas.

Moving a Lateral

A common problem with sprinkler irrigation is the large labor force needed to move the pipes and sprinklers around the field. In some places such labor may not be available and may also be costly. To overcome this problem many mobile systems have been developed such as the hose reel raingun and the center pivot.

Another system which does not need a large labour force is the drag-hose sprinkler system. Main and laterals are buried PVC pipes: one lateral covers three positions. Sprinklers on risers carried by skids are attached to the laterals through hoses (similar to garden sprinklers). Only the skid with the sprinkler has to be moved from one position to another, which is an easy task.

Operating Sprinkler Systems

The main objective of a sprinkler system is to apply water as uniformly as possible to fill the root zone of the crop with water.

Wetting Patterns

The wetting pattern from a single rotary sprinkler is not very uniform. Normally the area wetted is circular. The heaviest wetting is close to the sprinkler. For good uniformity several sprinklers must be operated close together so that their patterns overlap. For good uniformity the overlap should be at least 65% of the wetted diameter. This determines the maximum spacing between sprinklers.

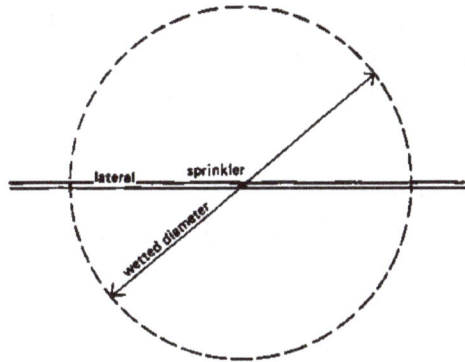

Wetting pattern for a single sprinkler

Wetting pattern for a single sprinkler

The uniformity of sprinkler applications can be affected by wind and water pressure.

Spray from sprinklers is easily blown about by even a gentle breeze and this can seriously reduce uniformity. To reduce the effects of wind the sprinklers can be positioned more closely together.

Sprinklers will only work well at the right operating pressure recommended by the manufacturer. If the pressure is above or below this then the distribution will be affected. The most common problem is when the pressure is too low. This happens when pumps and pipes wear. Friction increases and so pressure at the sprinkler reduces. The result is that the water jet does not break up and all the water tends to fall in one area towards the outside of the wetted circle. If the pressure is too high then the distribution will also be poor. A fine spray develops which falls close to the sprinkler.

Application Rate

This is the average rate at which water is sprayed onto the crops and is measured in mm/hour. The application rate depends on the size of sprinkler nozzles, the operating pressure and the distance between sprinklers. When selecting a sprinkler system it is important to make sure that the average application rate is less than the basic infiltration rate of the soil. In this way all the water applied will be readily absorbed by the soil and there should be no runoff.

Sprinkler Drop Sizes

As water sprays from a sprinkler it breaks up into small drops between 0.5 and 4.0 mm in size. The small drops fall close to the sprinkler whereas the larger ones fall close to the edge of the wetted circle. Large drops can damage delicate crops and soils and so in such conditions it is best to use the smaller sprinklers.

Drop size is also controlled by pressure and nozzle size. When the pressure is low, drops tend to be much larger as the water jet does not break up easily. So to avoid crop and soil damage use small diameter nozzles operating at or above the normal recommended operating pressure.

Advantages of Sprinkler Irrigation

- Elimination of the channels for conveyance, therefore no conveyance loss.
- Suitable to all types of soil except heavy clay.
- Suitable for irrigating crops where the plant population per unit area is very high. It is most suitable for oil seeds and other cereal and vegetable crops.
- Water saving.
- Closer control of water application convenient for giving light and frequent irrigation and higher water application efficiency.
- Increase in yield.
- Mobility of system.
- May also be used for undulating area.
- Saves land as no bunds etc. are required.
- Influences greater conducive microclimate.

- Areas located at a higher elevation than the source can be irrigated.

- Possibility of using soluble fertilizers and chemicals.

- Less problem of clogging of sprinkler nozzles due to sediment laden water.

Subsurface Textile Irrigation

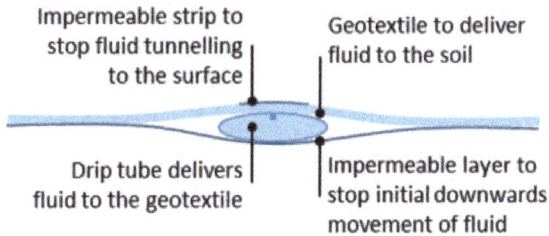

Impermeable strip to
stop fluid tunnelling
to the surface

Geotextile to deliver
fluid to the soil

Drip tube delivers
fluid to the geotextile

Impermeable layer to
stop initial downwards
movement of fluid

The structure of an example SSTI installation.

Subsurface Textile Irrigation (SSTI) is a technology designed specifically for subsurface irrigation in all soil textures from desert sands to heavy clays. Use of SSTI will significantly reduce the usage of water, fertilizer and herbicide. It will lower on-going operational costs and, if maintained properly, will last for decades. By delivering water and nutrients directly to the root zone, plants are healthier and have a far greater yield.

It is the only irrigation system that can safely use recycled water or treated water without expensive "polishing" treatment because water never reaches the surface.

A typical subsurface textile irrigation system has an impermeable base layer (usually polyethylene or polypropylene), a drip line running along that base, a layer of geotextile on top of the drip line and, finally, a narrow impermeable layer on top of the geotextile.

Unlike standard drip irrigation, the spacing of emitters in the drip pipe is not critical as the geotextile moves the water along the fabric up to 2m from the dripper.

SSTI is installed 15–20 cm below the surface for residential/commercial applications and 30–50 cm for agricultural applications.

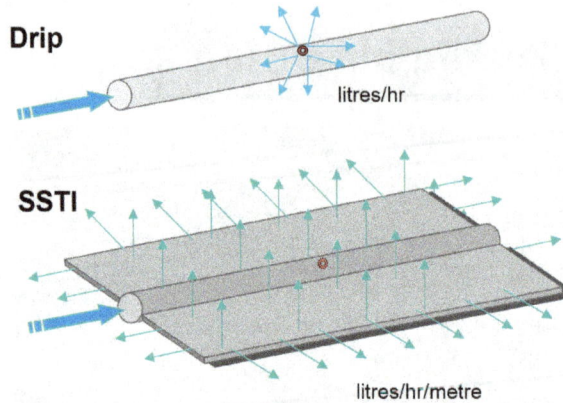

The flow of water through an SSTI installation as compared to a drip irrigation system.

To increase effectiveness, SSTI products should have an impermeable base layer to slow gravitational loss of water and to create an elliptical wetting pattern under the soil surface. It should also have a small impermeable top layer to ensure that water from the dripper does not "tunnel" through the geotextile and up to the surface (again a common problem with bare subsurface drip pipe). The effect of these two layers is dramatic as it maximizes the spread of water through the geotextile.

When comparing SSTI with surface drip, using the same amount of water, SSTI can cover 2.5 times the volume of soil and takes six times longer to dry down until the next irrigation is required.

Recycled Water and Treated Effluent

Recycled water can be used in SSTI systems as it will spread the nutrient load over 2-3 times the soil volume (compared to other irrigation methods). This means that additional nutrient requirements are minimized and the soil will have a long life without overloading other nutrients (especially phosphorus and potassium).

A major benefit of SSTI is that treated effluent can be used but is prevented from reaching the surface. Recreational or agricultural activities can continue on the field during irrigation without the contaminants coming in contact with the public.

Nutrients can be injected through all SSTI systems (fertigation). Macro and micro nutrients can be delivered to specific crops including grass, pasture, trees and vines. The nutrient is placed directly in the root zone so there is almost no wastage and no potential for run off into waterways.

Use of Geotextiles

Geotextiles were used for capillary irrigation not long after World War II. However, it was not until 1995 that extensive research was conducted at the CSIRO Division of Land and Water in Griffith, Australia, by Grain Security Foundation Ltd, that SSTI was established as a serious commercial alternative to drip.

Polyester geotextile of specific thickness and manufacture is required to ensure the system has the appropriate flow characteristics and so that it does not become hydrophobic (repels water).

Solving Typical Subsurface Problems

Studies were done on many forms of geotextile using various dripper rates and configurations to evaluate water flow for the major soil texture types and to determine if SSTI had the same problems experienced by subsurface drip systems (SDI). Specifically, considerable focus was placed on the problem of root intrusion, tunneling, dripper blockage and insect damage.

Root Intrusion

Research data showed that roots could invade the geotextile but did not thrive nor thicken in the polyester geotextile and, therefore, caused no damage. The roots were deterred from entering the dripper area because that area is dry relative to the rest of the system. Roots simply went elsewhere.

Tunneling

Tunneling (the process whereby water works its way up to the surface) was nearly eliminated by the narrow reflective impermeable tape above the drip line.

Blocked Drip Emitters

Drippers are often blocked in bare drip systems due to soil lodging in the emitters and/ or soil crusting around the emitters when the system is not operating. SSTI eliminates this problem because there is a physical barrier presented by the geotextile and due to the fact that the soil remains moist for much longer than drip systems (i.e. the soil does not form a crust anyway).

Insects

Similarly, insects are deterred from damaging SSTI systems because of the geotextile barrier.

Water Delivery

Initially solid, thick-walled drip pipes were used in trials at the CSIRO but subsequent trials between 1995 and 1998 demonstrated that drip tapes (thin-walled, flexible pipe) could be used very effectively. The loss of strength from using thin-walled drip tapes was countered by the additional tensile strength of the entire system (base layer, drip tape and geotextile). While significantly reducing the cost of SSTI, the use of drip tapes also permitted the use of large diameter drip lines (of up to 35mm) allowing for run lengths up to one kilometre. The thin, flexible wall of drip tape also meant that master rolls of manufactured product could be much larger (over 600m in length).

Optimal Width and Wetting Patterns

The width of SSTI products varies based on the application. However, at the University of Queensland, Australia trials were conducted to discover the optimal width of the SSTI in black cracking clay with alfalfa sown across a single SSTI line. The SSTI line was 20 cm wide and was installed 30 cm below the surface. The alfalfa germinated directly above the SSTI line but covered 75 cm either side (a total of 1.5 m) in uniform lines. As a result of this research, most SSTI systems are between 6 cm and 20 cm wide. However, some products are up to 2 m wide.

Installation Technologies

In 1997 the first SSTI plow was tested at CSIRO Griffith using a standard three point linkage behind a tractor. This demonstrated that SSTI could be installed at 20–40 cm deep quickly and easily using the plow. SSTI plows are now available to install 1, 2 or 3 lines (more lines are possible).

Other uses of SSTI

In 1996 the first ebb and flow mat technology for potted plants was commercialized based on SSTI technology using drip tapes to control the water delivery. This ebb and flow mat form of SSTI proved very effective in producing potted plants, sprouted wheat and barley for animal production and for research purposes in producing seed varieties without the use of any overhead irrigation. It also provided an effective fertilizer delivery platform.

Components

Well-designed SSTI systems are laid out essentially the same as drip systems but in many cases SSTI can be laid in a "serpentine" pattern vastly reducing the number of take-off connections and potential leaks.

Most drip tubes/tapes used in SSTI are pressure compensating. For example, using a 16mm drip tape, a run of up to 180 m can be achieved from one connection. Longer runs of up to 1,000 m can be achieved using lower flow rates per lineal meter and/or larger diameter drip tape.

The following components make up a typical SSTI installation:

- Pump or pressurized water source to 100-300kPa (14-43psi)

- Water filters or filtration system from 120 micron with suspended solids less than 30ppm

- Fertigation injector systems

- Back flow prevention

- Pressure regulating valves

- Main line. Can be LDPE or PVC

- Solenoid valves or gate valves to control water flow

- SSTI system laterals

- Barbed or Spinlock fittings with stainless steel clips

- Flushing valves at the end of laterals or combined laterals into a flushing line so that regular flushing can remove suspended solids or bacteria that may build up when using recycled water

Using drip tapes in SSTI means that there is a wide range of commercial fittings making SSTI very easy to install. Fittings are usually spinlock or ring lock devices securing the drip tape using a barb.

Advantages of SSTI

The advantages of SSTI are:

- SSTI is a "permanent" solution if maintained properly. The components are inert and, given that they are situated below the ground, are not subject to the effects of weather, animals, machinery, vandals or other terrestrial conditions.

- Water savings of 50-75% compared with overhead systems.

- Low pressure requirement (also means lower power requirements).

- Yields can be improved up to four times in certain crops.

- Minimal root intrusion to drippers in the SSTI with a deflective tape on top.

- No emitter blockage due to crusting.

- Minimal effect of evaporation.

- Safe use of recycled or treated water.

- Can use the field (for recreation or agriculture) while irrigation is running.

- Weed growth is minimized because the water does not reach the surface (saving herbicide cost). Germination of weeds only occurs during rainfall.

- Fertigation can be done directly to the root zone (saving fertilizer cost).

- Efficient distribution of nutrients to the entire root zone.

- Broad wetting patterns (moisture covers the entire root zone).

- Water delivery can match the natural capillary rates in soil so saturation is minimized.

- Soil moisture can maintained at field capacity (minimised gravitational losses).

- No surface run off.

- Soil erosion is minimized.

- Fields do not have to be perfectly level.

- Fields with irregular shapes can be accommodated.

- Distance between lines of SSTI is far greater than drip (lower number of solenoids and other components compared to sprinklers and drip).

- SSTI laterals can be ploughed in behind a tractor on large sites (over 10,000 m per day).

- Foliage remains dry (fungal and bacterial leaf disease is minimized).

Disadvantages of SSTI

The disadvantages of SSTI are:

- Initial capital cost is typically more than overhead irrigation.

- Quality installation is critical. If mistakes made, they are difficult to find.

- Installation using correct fittings must be done.

- Regular maintenance is required to ensure long life.

- Automated control and monitoring systems are preferable (subsurface irrigation does not give any visual indicators to show if it is working or not).

- SSTI is usually not UV-treated so it must be kept out of sunlight until it is installed under the surface.

- Temporary overhead watering may be required to establish turf in hot areas.

- Germination of some agricultural crops may require overhead watering if insufficient rainfall.

- SSTI cannot apply fertilizers or herbicides overhead on the surface.

- Rodents may damage the system (although less than drip systems).

Center Pivot Irrigation

As the name suggests, center pivots irrigate in a circular pattern around a central pivot point. Pivots are capable of applying water, fertilizer, chemicals, and herbicides. This versatility can improve the efficiency of irrigation practices by using a single piece of machinery to perform several functions.

A center pivot or lateral move basically consists of pipeline (lateral) mounted on motorized structures (towers) with wheels for locomotion. A center pivot machine rotates around a " pivot" point in the center of the field whereas a lateral move machine travels along a straight path and has a separate guidance system. Sprinkler outlets are installed on the top a pipe supported by steel trusses between adjacent tower structures. The towers are usually 90 to 200 ft (30 to 60 m) apart and each tower has a 1 hp motor and sits on two large rubber or steel tires. The combination of pipe, truss and sprinklers between two towers is called a span. Flexible couplers at each tower connect the pipes of two adjacent spans. The maximum length of span is a function of pipe size, pipe thickness (strength), field slope and topography. Span length does not have to be uniform; in fact, it is often varied to match field dimensions or to provide adequate clearance between the truss and soil surface on rolling terrain. An "overhang" is a smaller pipe with sprinklers that is often suspended by cables beyond the outermost tower(s) to increase the w etted area. Large volume end guns and corner systems may also be added to the end of a machine to increase the wetted area in the corners or to cover additional areas. Machines can be more than 4900 feet (1000 m) long although the most common length of a basic machine is about 1300 ft (~400 m). The distance between the trusses and the ground can range from 4 to 14 ft (1.2 to 4.3 m) with most between 8 and 9 ft (2.5 and 2.8 m). Basic system life should be 15 to 20 years not including sprinkler heads, pumps and other ancillary equipment.

Most machines are powered by electricity, although some manufacturers use hydraulic motors, which are more expensive. A one constant speed horsepower electric or hydraulic drive motor is used to propel each tower. Tower motors should always be covered to extend their useful life. Electric power wires and/or hydraulic lines run the length of the machine with control boxes or valves at each tower. A primary control panel is usually located at the pivot base or at the engine on lateral moves. Hydraulic powered systems have a higher initial cost but may have lower annual costs because of lower maintenance and operational costs. Various manufacturers are looking at using more expensive variable speed electric motors to reduce start-stop effects on uniformity, especially with chemigation systems.

Not including a pivot base structure, there can be one to fifteen or more towers on each system. Towers are usually identified by number starting with the tower closest to the pivot base or the linear move engine/pump assembly. The towers should always follow the same tracks through the field. Equipment crossings and problems with traction as well as runoff from the compacted, wet wheel tracks are sometimes serious concerns.

Keeping linear move and center pivot machines in good alignment is critical to proper operation. Substantial damage can occur to the equipment and the crops if the alignment system fails. Alignment sensors are located on the pipeline at each tower causing the tower motor to start or stop. Alignment is controlled by electronic strain gauges, radio or laser controlled sensor systems. Often the first tower (closest to primary controls) will have an additional adjustable timer that will turn off the entire system if that tower does not move at least once every two to five minutes for extra protection against alignment system failures. Inadequate traction at a tower will often cause alignment problems.

Generally, the tower farthest from the pivot point controls the movement of the entire machine. Minimum rotation times (maximum speed) are commonly between 14 and 20 hours (2 to 3 m/min at the outer tower). Special high speed gear boxes can be installed on each tower to reduce rotation times to less than 12 hours (i.e., 4.3 m/min at the outer tower) which is often desirable on sandy or cracking clay soils. Timing controls at the control panel determine the relative average speed of the outer tower. The on-off cycle time of the outer control tower is usually about 1 minute (i.e., on a 50% speed setting, the outer motor is on 30 seconds in every minute.) A 100% setting causes the machine to travel at maximum speed (minimum rotation time) whereas a 50% setting results in the outer tower moving at half the maximum speed. Of course, the slower the rotation speed, the greater the amount of water applied. All the other towers try to stay in alignment with the end tower as regulated by the alignment system. However, tower movement in the interior of the system is somewhat random and start and stop times of 1 to 3 minutes may occur. Thus, because of the start stop action the uniformity coefficients in the direction of travel are largest near the pivot and the end tower and the smallest near the center of the irrigation system.

Since all the towers on standard electric systems typically have the same motors and gearing ratios, the towers start and stop to stay in alignment. The start-stop action of the towers is not usually a problem with water application uniformity, which is averaged over several days. However, with pesticides becoming more specific, applied in small amounts and costing several dollars per gram, this jerky movement may present a concern if pesticides are to be applied through the irrigation water or if a separate system attached to the center pivot trusses. Costly variable speed electric motors on each tower may be justified for pesticide systems where uniformity is critical. Hydraulically powered tower motors can be adjusted so that there is no start-stop movement and applications are more uniform.

Flow requirements for center pivots and lateral move systems are defined in terms of total flow supplied to the system or the total system capacity, Q_T (gpm), for hydraulic requirements, and gross system capacity, Q_g (gpm/ac), for irrigation requirements and management. Q_g is defined as the total system capacity, QT, divided by the total irrigated area. Standard inside diameters of center pivot pipes are 5.31, 5.74, 6.38, 7.76, 8.37 and 9.76 inches (135 mm, 146 mm, 162 mm, 197 mm, 212.7 mm and 247.8 mm, respectively) with most (50-52 hectare)125 -130 acre systems using 6.38 inch (162 mm) steel tubing with a wall thickness of 0.11 inch (2.77 mm).

The two types of sprinklers used with center pivot and linear move systems are "impact" and " spray" heads. Impacts (traditional, older style impact drive) are generally low pressure, low angle (6° to 15°) heads mounted directly on the top of the pivot lateral pipe. Spray heads are subdivided into sprayers and "rotators." Sprays provide a mist or small jets that can also be mounted on top of the pipe but are more commonly installed at the bottom end of flexible drop tube connected to a U-shaped "gooseneck" on the lateral pipe. The height of the sprinklers may be adjusted throughout the season to maintain them above the crop canopy but this may negatively affect uniformity. The height, location, spacing, size and the discharge from each head are specified in the sprinkler "package" from the manufacturer. A standard 1300 ft (400 m) long center pivot will have 100 to 110 sprinklers. Low pressure spray type sprinkler heads mounted close to the canopy are probably the most popular to reduce wind and evaporation losses although low pressure impact heads on the pivot lateral are still used in some areas. Use of high pressure impact heads is becoming rare. Use of pressure regulators or flow control nozzles is very common with low pressure systems.

Water losses from spray heads near the top of the canopy typically range from 0-2% due to droplet evaporation, wind drift is usually less than 5%, evaporation from crop canopy ranges from 4 to 8%, and soil evaporation less than 2% whereas runoff may range from 0 to 15% or more depending on slope and soil conditions. Spray heads and impacts mounted on top of the pipe lateral may have droplet evaporation and w ind drift losses as high as 15%. Evaporation may slightly offset crop water use, but this amount is difficult to measure or calculate and is usually less than 15% of total ET. For spray irrigation on drops over a crop with a full canopy, application efficiencies of about 90 to 92% are attainable with no surface runoff whereas sprinklers on the top of the pipe may attain efficiencies from 80 -85% .

The first center pivots had sprinkler spacing's of about 32 ft (9.75 m) using impact heads. Later versions had variable spacing's with sprinklers closer together as they approached the distal end of the lateral. Most modern machines have a constant outlet spacing ranging from 3 to 9 ft (1 to 3 m) depending on the manufacturer and the type of system. Machines can be ordered for almost any outlet spacing although 6 to 8 ft (2 to 2.4 m) spacing's are typical, although LEPA machines may have outlets as close as every 2 to 3 feet (0.6 to 1 m). Sprinklers are installed in every outlet for linear move machines. How ever, on center pivots near the pivot where machine movement is slow, not every outlet has a sprinkler installed in order to reduce application depths. Generally, after the first tower all outlets have sprinklers installed. Uniform water applications depend on the careful matching of spacing, the particular sprinkler heads to be used and their height above the crop canopy.

Older machines and sprinkler packages often required pressures in excess of 70 psi (500 kPa) to operate, but modern machines are generally designed to operate at 35 psi (250 kPa) or less. These pressures are often too low for high volume-high pressure end guns and small electric booster pumps are often installed at the last tower. Discharge from end guns and corner systems should be controlled to avoid water applications to roads, streams or drainage facilities (especially when chemicals are being applied) by switches at the pivot base.

Pipe sizes range from about 4 inches (100 mm) to more than 12 inches (300 mm) in diameter. Pipe sizes on center pivots may decrease along the length of the lateral with increasing distance from the pivot. Linear move systems have uniform pipe sizes except for overhangs past the end towers. However, minimum pipe size is determined by strength rather than hydraulic considerations.

Regardless of the type of system, strict annual or weekly maintenance and lubrication schedules are required for all motors, gear boxes, alignment systems, couplers, seals and other parts. Sprinkler heads and pressure regulators should be replaced on a regular basis, usually every 3 to 4 years.

Lateral Move Systems

A lateral move machine travels in a straight line to irrigate up to 95% of square or rectangular shaped fields, and is supplied water through a carefully constructed open ditch that runs parallel to the direction of travel or large flexible hose-buried pipeline system. Some land is lost to production because of area needed for the supply ditch or supply hose drag lanes (30-40 ft [10-15 m] wide for the length of the field). Generally, linear move systems are used on land with slopes less than 6%.

Many lateral move systems have large diesel engines connected to a generator that powers the pump (open ditch supply) and tower motors. A hose drag system is pressurized by pumps off the field. The engine/pump/control assembly can be located in the center of the lateral move or at the edge of the field. These systems are guided by buried electrical cables, lasers or wire on stakes. All the sprinklers on a lateral move are usually the same size except for large end guns on some systems. The design of a lateral move system is a simplified case of center pivots although the management and operation is more complex with higher labor requirements. The average intensity of application, I (in/hr), for lateral move systems is given by:

$$I = \frac{K * Q_r}{W * L}$$

where Q_T is the total water delivered to the system also called the total system capacity (gpm) , K is a units conversion factor equal to 96.25 in English units (K is 6.0 for metric units using l s-1 and m with I in mm/min), W is the width of the sprinkler pattern in feet, and L is the total length of the lateral pipe including overhangs in feet. The actual depth applied during an irrigation event for a lateral move system is given by:

$$d = \frac{K * Q_r}{L * S}$$

Where d is the applied depth in inches, K is a units conversion equal to 1.604 in English units (equals 1 .0 in metric with mm, m and m s-1) and S is the speed of movement of the machine in ft/min.

Lateral move systems are often used where productive land is limited and valuable. Capital costs of lateral move machines will vary from US$500 to over US$1200 per acre (US$1250 to over US$3000 per hectare) not including water source or land costs. Labor can be a significant annual cost factor.

Center Pivot Systems

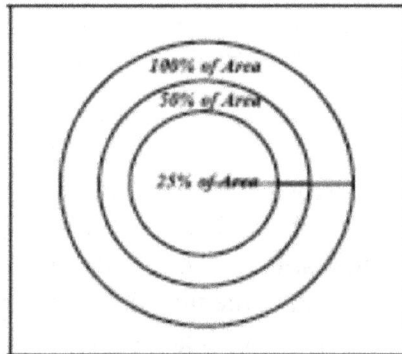

Percent of areas covered at various radii of a center pivot irrigation
system without end guns or corner systems

A center pivot machine rotates in a circle around a base pipe structure in the center of the field so that the irrigated section is any circular shape including parts of circles less than 360°. They can cover 80% to 90% of the area of a square field. Center pivots can operate on widely variable terrain with slopes as much as 30% with proper design although an upper limit of 15% slopes is generally recommended. A service road is usually necessary for control adjustment and maintenance as well as refilling, operation and monitoring of any chemigation supply tanks and injection pumps located at the pivot.

The pivot base structure is also the source of water, power and control wires. Pivot bases are usually permanent for large systems but may be portable for towable systems. Electrical power is supplied to tower motors, hydraulic and booster pumps through the slip ring connection at the pivot base.

The percent of area irrigated at various radii of a center pivot are illustrated in figure above where the innermost circle is at 50% of the radius but only contains 25% of the total irrigated area. It is important to realize that 75% of the total cropped area occurs in the outer half of the radius. Thus, management concerns tend to focus on outer towers, but many of the disease and water distribution problems will occur in the inner portions of the circle. The current state of the technology basically treats the entire field as a uniform soil and crop system. Some of the new control panels do allow changes in speed in selected sectors, but field variations are seldom pie shaped.

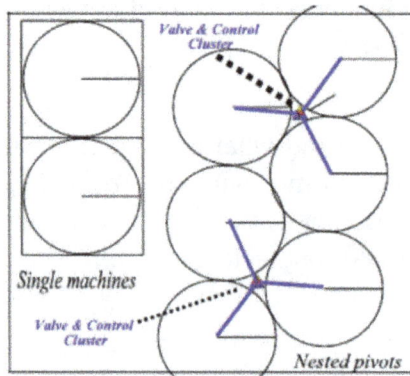

Examples if single machine placement and the increase in
cropped area achieved by nesting pivots

Center pivots are available to irrigate from 5 to 500 ac (2 to 200 ha) although a typical machine will generally irrigate about 125 to 130 acres (50-52 ha). Economic considerations usually limits their application to irrigated areas larger than 50 ac (20 ha). The area irrigated with a center pivot depends on the radius of the main lateral plus the radius increase due to end guns and corner system s. If the center pivot is positioned in the middle of a square piece of land without an end gun, the machine will irrigate about 80% of the total area. Machines are often nested (clustered) together if several center pivots are installed on one large piece of land so that 85% to 95% of the total area is irrigated.

The average operating pressure of a center pivot lateral will vary significantly depending on whether the pipeline is going up hill or down hill. This can result in large variations in sprinkler discharge so that pressure regulators or flow control nozzles are often required on every sprinkler head.

The area irrigated by a machine can be extended by the addition of relatively inexpensive high volume end guns and expensive "corner systems." Center pivot capital costs can range from US$400 to more than US$1000 per acre (US$1000-US$2500 per hectare) not including land and water development cost depending on options such as size, sprinkler packages, corner systems and end guns.

Corner Systems: Corner systems (also called corner "catchers" or swing spans) may be installed on center pivot systems to increase the irrigated, productive areas in the corners and other non-symmetrical regions along the field boundaries by adding 17-25 acres (7 to 10 ha) without buying or renting more land in areas where circles cannot be clustered. They usually consist of an additional tower and pipe system that is connected to the last tower of the main system. Corner systems normally follow behind the main system which tends to fix system rotation direction (rotation direction can be changed with great caution). The corner system tower generally has a guidance system such as phased-lock-loop circuitry that detects a low frequency radio signal from a guidance wire that is buried directly below where the tower will run. Signals are received by a high resolution antenna and receiver and fed into a microprocessor which continually monitors tracking and activates the steering motors. Sprinklers on the corner system are likewise controlled by the same microprocessor which also activates individual solenoid valves depending on their location with respect to the main end tower and the edge of the field. Use of corner systems on slopes greater than 15% may be problematic. Some corner systems use variable speed motors to improve water distribution and operation.

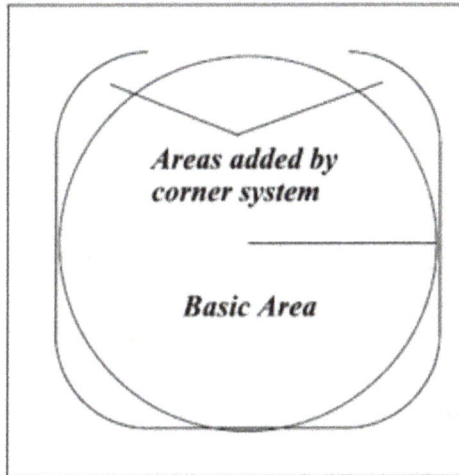

Areas added by corner system

Basic Area

Schematic showing additional land covered by a corner system compared to a standard machine

The incremental cost per hectare for the additional land covered by corner systems may be two to three times as expensive as land covered by the main system. Therefore, corner systems are usually only added when land values are high. In addition, a corner system can cause large fluctuations in demand from the water supply system as it turns sprinklers on and off resulting in large, undesirable pressure fluctuations and poor water distributions. Pressure and flow problems may require the installation special controls and valves or expensive variable speed pump motors. These corner system sections also tend have fairly high maintenance costs due to the complexity and many moving parts. In general, corner systems should be added only after careful analysis of all the economic benefits.

End Guns: End gun systems composed of one or more large, high pressure heads are often used to extend the area in the corners of single machines on square blocks of land. These are used as a low cost alternative to corner systems to expand irrigated area as much as 21% in corners over a 35° to 42° arc. Considering the amount of land added by a relatively small increase in radius, end guns are popular as an inexpensive way to add significant irrigated acreage to the field, however, they are not without problems. A major consideration with end guns is that they are a basically a single large sprinkler and the application depth tapers off with distance and severe drought stress may occur at the field edges. This may not be significant for biomass production (e.g., forage crops) but can be a major problem when deficit soil water conditions negatively affect crop quality. For this reason, many irrigators will turn off the end guns on their center pivots when potatoes are being grown. Booster pumps are usually required to operate the end guns which can result in the same pressure and flow problems experienced with corner systems.

Towable Systems: Towable systems can irrigate from one to four adjacent fields and are often smaller systems than fixed machines. These systems are typically used as supplemental irrigation in humid areas and used on more than one field when rainfall is insufficient for good crop production. In these cases, the machine is typically moved at least daily during the drought periods. They are often not practical in arid areas on more than one field at a time because of the high labor required for the frequent moving of them from field to field. In more arid areas the towable systems are usually moved once or twice a year as part of rotation program or on leased land.

Orchards and Vineyard Irrigation: Orchards and vineyards can be irrigated with center pivot and linear move machines. However, if the water cannot be applied below the canopy there are a number of cultural and fungal disease problems that may develop. Consequently, the fields must be designed and planted so that center pivot or linear move systems apply the water below the canopy using small sprinklers or bubblers similar to the LEPA systems. Vineyards can be irrigated with standard height machines. Orchards, on the other hand, have been successfully irrigated with high clearance (i.e., 4.3+ m high) machines.

Orchards and vineyards are generally planted in circular rows for center pivots and in straight rows for linear move systems (similar to the LEPA systems below). The irrigation machines are usually special orders to fit each specific installation. Sprinkler, spray or bubbler heads are suspended between row s on long drops with heads that apply water below the tree or vine canopy almost at ground level. As a water conservation measure and to ensure good coverage (and no cover crop between rows), there are typically two 180° flat or downward spray heads directed towards the plant row are located on each side of the plants. Fertilizers, systemic pesticides and pre-emergent herbicides can be applied with the irrigation water.

LEPA Systems

A special adaptation of the technology is the Low Energy Precision Application (LEPA) method that can be installed on both center pivot and linear move systems. LEPA has "drop" tubes about every meter that extend to the soil surface where a low-pressure bubbler is attached in place of a sprinkler. Water is applied directly to the furrow and evaporation losses are minimized since the canopy is not wetted. These systems can be very efficient (e.g., 95- 98%) since evaporation losses (soil evaporation generally less than 2% with alternate row irrigation, although runoff may be as much as 50 % with poorly designed and operated systems) are minimal although initial capital costs are higher than standard systems.

Crops are usually planted in a circle so that the drops do not damage plants. Sometimes a canvas "sock" or other fabric energy dissipation device is used to prevent soil erosion in the furrows. The use of a machine such as the Dammer-Diker™ is often used to create small reservoirs to store water until it has infiltrated on heavy or steeply sloping soils under both LEPA and regular application techniques. Typical quarter mile long (400 m) LEPA systems will have 350 to 450 heads.

Operational Characteristics of Center Pivot Irrigation Systems

Many consider the current center pivot technology to be mature. They are mechanically reliable, simple to operate and require little supervision. However, the management for these systems is much different and unique compared to other types of irrigation systems. These systems are inherently characterized by light, frequent irrigations (e.g., daily) which offers numerous water and nutrient management advantages as well as numerous cultural disadvantages.

From a benefits standpoint, water and water soluble nutrients can be carefully applied in amount to exactly meet plant needs. The light applications can potentially reduce leaching in sandy soils (or cracking clays). Culturally, the frequent wetting of the canopy often creates ideal conditions for many fungal diseases, especially inside the tower closest to the pivot base. Shallow root development is encouraged on many crops by the frequent, light irrigations so there is often little buffering against drought stresses in the event of a mechanical breakdown of the system. For this reason, soil water contents in the upper regions of the root zone generally have to maintained at relatively high levels.

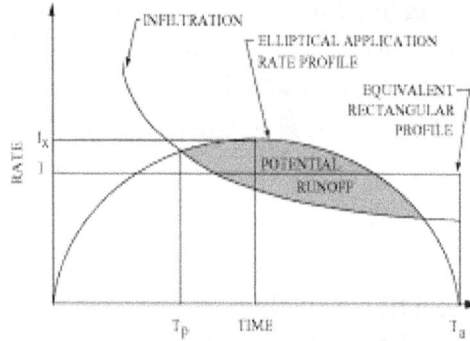

Interaction between the infiltration function and the
application rate pattern (Tp is time to ponding, Ta is the application time).

The frequent irrigations require adjustment of rotation times such that the machine is not in the same spot in the field every day at the same time to "average" losses and over applications across the field over time. Thus, rotation times that are 12 hour multiples are avoided.

Matching Application and Infiltration Rates

A major physical phenomena that center pivots take advantage of is that initial water infiltration rates into soils are high. Light, quick applications take maximum advantage of this phenomenon. To illustrate, the outermost tower of a basic 125+ ac (50+ ha) center pivot can travel 3 to 15 ft per minute (1 to 4 m min-1). However, the innermost tower travels only about 10% of that speed. This means that sprinklers at the outer tower are applying water 10 times faster than those near the first tower in order to have the same depth of application applied along the entire length of the pivot. With some sprinkler packages the application rate at the outer tower may exceed 4 in./hr (100 mm/hr).

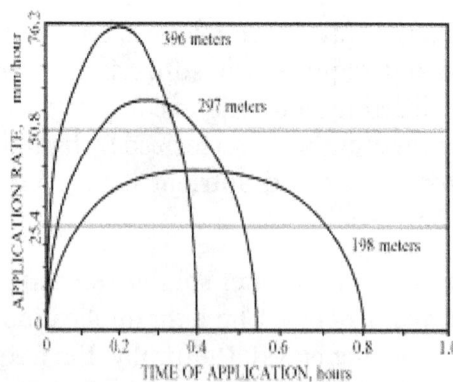

Application rate profiles and time of wetting showing the wetted diameters of the
sprinkler heads depending on their position along the lateral length relative to the
pivot base with a uniform width of wetted strip.

Thus, the sprinklers at the end of the machine generally cover larger diameters even at higher system rotation speeds to avoid exceeding the infiltration rate of the soils. Figure below shows the interaction between the application rate and the infiltration function. The intensity of application is illustrated in figure below relative to position in the field assuming the same amount of water is being applied showing the different wetting times. The rate of water application reaches a peak when the sprinkler passes directly over a location at a, b and the full radius of a basic system. The objective of proper nozzle selection and system operation is to ensure that the application rates do not exceed the respective infiltration rates at various points along the lateral.

In order to meet the application depth requirements, the discharge from sprinklers basically increases linearly with the radius as shown in figure above. In addition, the majority of the pressure loss occurs in the first one-third of the lateral pipe. A sprinkler at 985 feet (300 meters) from the pivot base will have twice the discharge of a sprinkler at 492 feet (150 m). Thus, the required discharge for any individual sprinkler along the pivot lateral is:

$$q_s = \frac{K\pi RSQ_g}{K}$$

where q_s is the individual sprinkler discharge in gpm, R is the radial distance from the pivot base in feet, S is the spacing between adjacent sprinklers along the pipe lateral in feet, and Qg is the gross system capacity for the irrigation system in gpm per acre, K is a units conversion of 43560.0 in English units (equals 10000. 0 in metric units of L s-1, m, and L s-1ha-1). Special soil cultural practices may have to be implemented if the system capacity results in runoff in various areas of the field. Table below presents sprinkler discharge requirements with length as a function of gross system capacity.

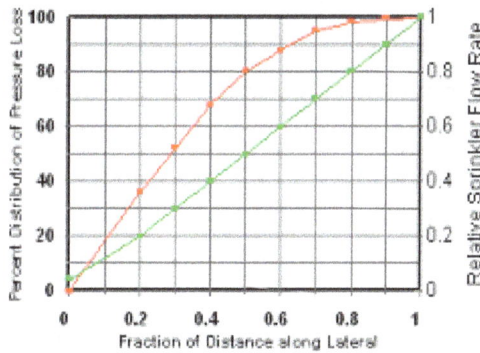

Typical distributions of pressures (upper line) and flow rates (lower line)
from individual sprinklers along the lateral length

Assuming uniform crop, soil, microclimate and topographic conditions, the goal of irrigation is to have the most uniform water application pattern possible. Two major variables with selection of sprinklers are spacing and the type or size of the heads. Sprinklers must be spaced close enough to have good overlapping of wetting patterns. The sprinkler type and discharge must be selected to avoid runoff. They must be matched to the soil and crop. The kinetic energy and power with which droplets impact the soil can have a large effect on soil compaction and sealing which can greatly increase runoff (a major problem with end guns), and small droplets may be beneficial in reducing soil sealing. However, small droplets are subject to wind drift and evaporative losses. Thus, selection of best sprinkler heads is, at best, a compromise between often-conflicting criteria and additional measures such as creating small storage reservoirs in the furrows may be required.

Generally, nozzle sizes are small near the pivot base and gradually increase in size and discharge as the radius increases. Corner systems generally have similar or slightly larger nozzles as the end tower of the basic system. On very long systems, the largest nozzles may not have enough flow capacity and two or more sprinklers must be installed or the system flow requirements reduced and system pressure increased. Infiltration problems are often reduced by spreading water applications over are larger area through placement of drop tubes on alternating sides of the pivot lateral truss structure or having two to four small heads on a small boom mounted almost perpendicular to the pivot lateral.

Radius, ft	System Capacities (gpm/ac)					
	4.0	5.0	6.0	7.0	9.0	10.0
300	0.17	0.22	0.26	0.30	0.39	0.43
600	0.35	0.43	0.52	0.61	0.78	0.87
900	0.52	0.65	0.78	0.91	1.17	1.30
1200	0.69	0.87	1.04	1.21	1.56	1.73
1500	0.87	1.08	1.30	1.51	1.95	2.16
1800	1.04	1.30	1.56	1.82	2.34	2.60
2100	1.21	1.51	1.82	2.12	2.73	3.03
2400	1.38	1.73	2.08	2.42	3.12	3.46

Sprinkler discharge requirements in L/s per meter (1 L s-¹ha-¹ equals 6 .414 US gallons per acre).

Radius, ft	System Capacities (L s^{-1}ha^{-1})					
	0.6	0.8	1.0	1.2	1.4	1.6
100	0.038	0.050	0.063	0.075	0.088	0.101
200	0.075	0.101	0.126	0.151	0.176	0.201
300	0.113	0.151	0.188	0.226	0.264	0.302
400	0.151	0.201	0.251	0.302	0.352	0.402
500	0.188	0.251	0.314	0.377	0.440	0.503
600	0.226	0.302	0.377	0.452	0.528	0.603
700	0.264	0.352	0.440	0.528	0.616	0.704
800	0.302	0.402	0.503	0.603	0.704	0.804

Soil infiltration functions must be determined prior to the design and nozzle selection process and related to opportunity time. The opportunity time is less towards the outer towers since the sprinklers are traveling faster which must be offset by higher flow rates and wider wetted diameters per head with increasing distance from the pivot base. Thus, to have the same application depth, sprinkler selection and rotation speed are dependent on bare soil (no crop) and topographic factors to avoid soil erosion and wasting water and energy. The speed, S_j, at point j on the radius r is $2_j / t_{rotation}$ where $t_{rotation}$ is the time required for a complete rotation in minutes. The infiltration opportunity time t_j is defined as the wetted diameter of the nozzle, W_j, divided the speed, S_j. The average intensity (application rate in./hr), I_j, of application at radius (r) at point j which can be calculated as:

$$I_j = \frac{d_g}{t_j} = \frac{K r_j\, Q_g\, R_e}{W_j\, L^2}$$

where K is a units conversion factor equal to 192 .3 for English units (7200.6 in metric units of l s^{-1} and m), Q is the gross system capacity for a basic circle (no end guns or corner systems) in gpm, L is the radius of the basic circle in feet and W_j is the width in feet of the sprinkler application pattern at j. R_e is a loss factor equal to 1.0-fraction of estimated evaporation and wind drift losses (e.g., 0.10 to 0.15). Thus, this equation must be solved in a trial-and- error procedure until the selected nozzles and minimum system rotation speed do not cause runoff. Use 50% or less of the maximum system rotational speed for these calculations to allow for changing conditions through the season and management flexibility. Thus, the average application rate, I_L, in in/hr at the end of the basic pivot, L, is:

$$I_L = K \frac{L d_g}{W\, t_{rotation}}$$

where $t_{rotation}$ is in hours, dg is the gross daily application depth in inches, L and W are in feet, and K is a units factor of 6.2832 (same in metric units of mm/hr and meters). The average depth of water applied, d_g, per revolution is given by:

$$d_g = K\, Q_g\, h_r$$

where K is the units conversion equal to 0.00221 (K is 0 .36 in metric units of mm and l s-^{1}ha-^{1}), Q_g system capacity in gpm/ac, and hr is the hours per revolution.

The speed control setting, C_s, is a ratio of the depths at 100% speed (minimum depth per revolution) to the desired depth at a lower speed. It can also be determined as the ratio of the depth per revolution at the 100 % speed, d_g, to the application depth per revolution at a given speed, S, in percent, which is:

$$C_s = K \frac{R_t * v_m * d_g}{Q_r}$$

where K is a units factor equal to 31.168 (0.84 in metric units of l s-^{1}, meters, meters per minute and mm per revolution), Q_T is the total system capacity in gpm, R_1 is the distance from the pivot base to the end tower in feet, $_m$ is the velocity of the end tower in feet per minute for that revolution, and d_g is in inches applied at the 100% speed setting that should not result in runoff . Substituting equation 6 into equation 7 gives C_s in terms of velocity and hours per revolution (K would be 0 .01667 in both English and metric units).

System Selection Considerations

Because center pivot technology is so well developed, it may seem that design concerns are minimal. However, this is absolutely not the case. The machines must be designed to match each site. Information must be collected to characterize the variability of the soils (physical and chemical), topography, infiltration rates, microclimates across a field, and expected crop water use patterns over the season. Water supply (quality, quantity, timing and long term availability) must be investigated as well as any other potential physical, legal or social constraints must be identified. Steepness of slopes along wheel tracks can affect performance and long system life. Potential losses such as wind drift, evaporation, runoff, and deep percolation need to be estimated. Sprinkler spray and distribution patterns characteristics must match the soil. Machine operational criteria and strategies must be developed.

Total system flow requirement is needed for hydraulic calculations and proper selection of pumps, main lines and lateral sizes. The calculation for total system flow, Q_T (gpm), is:

$$Q_T = K * A * d_g$$

where K is a units factor of 18.86 (0.1158 in metric units of L s^{-1}, hectares and mm), A is the total area irrigated in acres and d_g is the daily gross depth applied per unit area in inches over the irrigation period of 24 hours. The total system flow, Q_T, can also be calculated based on the amount of water to be applied over a fixed time period, f, as:

$$Q_T = K \frac{A * d}{f * T}$$

where f is in days, d is the average depth applied in inches, T is hours per day of operation, and K is a units factor equal to 452.6 (use 2.78 in metric units of l s^{-1}, hectares and mm).

The gross system capacity, Q_g, is a useful quantity for easy comparisons of the adequacy of design and management of different sizes of center pivots and lateral move systems. It should be the amount of water continuously delivered to the machine per acre that is sufficient to meet peak evapotranspiration (ET) requirements as w ell as losses. Q_g typically ranges from 4 gpm/ac (0.65 l s^{-1} ha^{-1}) to as much as 10 gpm/ ac (1.6 l s^{-1}ha^{-1}) although the average is around 6 to 8 gpm/ac (1.0 to 1.2 l s^{-1}ha^{-1}) which allows some flexibility in case of breakdowns. Some systems are specifically designed to operate at a deficit during the peak water user periods to stretch water supplies or add additional center pivots to increase the total irrigated area, but this should be done only after careful consideration of all factors. A properly selected Q_g should account for local crop, climate and soil characteristics.

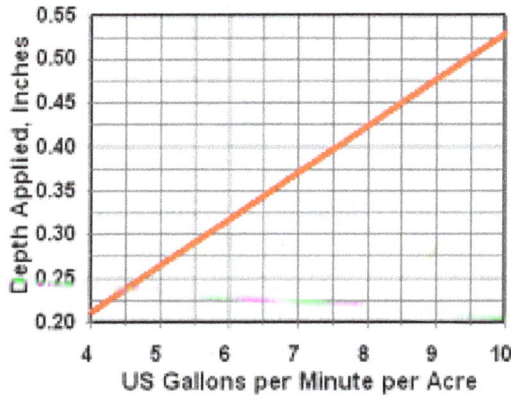

Water application depths in a 24 hour period for center pivots as related to the gross system capacity with no losses (100 % efficiency).

Some systems are deliberately under designed (gross system capacity too small) in order to stretch water supplies and irrigate more area, realizing that the plants may be stressed during peak water use requirement periods. Designing for less than 6 gpm/ac ($\sim$1 L s^{-1}ha^{-1}) is not recommended in arid areas of the Pacific Northwest since many crops would be stressed during peak water use periods. A design gross system capacity value of 8 gpm/ac (1.2 L s^{-1}ha^{-1}) is more appropriate.

Irrigation Scheduling

Center pivots and lateral move sprinkler systems will usually use more water than other sprinkler methods because the frequent irrigations have more evaporation losses from the plant canopy and soil as well as wind drift which occur with every application rather than once every 7 to 10 days. LEPA systems will be more efficient. Irrigations should be scheduled based on soil water levels to avoid undesirable levels of crop stress. This is compounded by the light frequent applications, shallow rooting and cultural operations such as fertigation, spray and tillage programs. If system capacity is not adequate to meet peak water use requirements, it may be necessary to build up soil water reserves and encourage deeper rooting prior to any water short periods, however, this it is often difficult to limit over-watering and avoid deep percolation losses.

Pressure variations due to static topographic differences and changes in system flow rate by the operation of end guns or corner systems can easily result in greater than 10 % variations ($\pm$5%) in water applications across a field. These differences in application depths are small on a daily basis, but the effects are additive over the season and may have serious consequences on water sensitive crops such as potatoes. Thus, the actual daily variation in application depths and the actual distribution of water must be considered in irrigation scheduling. General irrigation scheduling concerns are presented in the adjacent box.

Water Requirements. Center pivot and linear move systems inherently provide frequent, small water applications. Consequently, the volume of water stored in the soil and available for crop use can be considerably less than the wetted soil volume under other types of sprinkler irrigation. However, this practice can maintain higher, less variable soil water contents than other irrigation methods and reduce the occurrence of plant water stresses if the system capacity and management are appropriate.

The basic philosophy of center pivot and linear move systems is to replace water in the root zone in small increments as it is used by a plant at frequent intervals rather than refilling a much larger soil water reservoir after several days or weeks. Thus, the major concern for scheduling center pivot systems is primarily how much to apply during irrigation since the irrigation interval is often fixed by other factors, including design. The actual seasonal water requirements of a crop can be obtained from various technical sources including the Cooperative Extension Service.

The gross required depth, dg, in inches (in metric use mm per day of ET and P) of application per day is given by:

$$d_g = \frac{k_f\, ET_c - P_e}{E_a\,/100}$$

where k_f is a frequency factor to adjust the actual daily crop evapotranspiration (Et_c) in inches during the period for high frequency water applications which is typically about 1.2 for daily irrigations, 1.1 for irrigations every two days and 1.0 when irrigations are every 5 days or more. ET_c can be calculated from the Modified Penman, Penman-Monteith or other suitable estimating equation depending on data availability. This adjustment is necessary because of increased evaporation losses from the plant canopy and soil surface inherent in high frequency irrigation regimes. P_e is the effective precipitation during the period in inches and E_a is the application efficiency as a percent.

Using the same variables as previously defined except that d is the net depth applied, the maximum time between

$$f = \frac{d}{k_f\, ET_c - P}$$

irrigations, f, can be calculated as:

The estimated crop water use (ET) combined with amount of the area to be irrigated, will determine the total volume of water to be applied in an irrigation subject to system capacity. The system should be able to apply the maximum depth of water needed during peak water use periods accounting for all losses. The maximum interval between irrigations is primarily controlled by soil hydraulic characteristics, soil profile layering, and maximum allowable deficit levels for the crop. The depth of root zone soil, saturated hydraulic conductivities and soil water holding capacities may the volume applied in a single irrigation to avoid runoff or excessive deep percolation.

It is sometimes not possible to achieve optimum irrigation schedules because of irrigation system limitations. These include inflexibility in controls and instrumentation, inadequate system hydraulic capacities (including fill times and system drainage), and the quantity and quality of available water throughout the season.

Management considerations such as the quality and quantity of available labor can affect the ability to implement irrigation schedules. Timing, amount and label requirements for chemigation may also negatively influence optimum schedules. Irrigation schedules may have to be adjusted because of cultural or harvesting considerations or to take advantage of lower "off peak" electrical power rates.

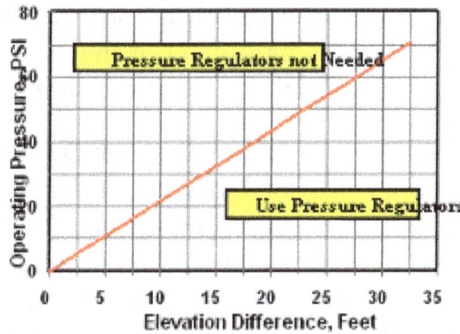

Suggested recommendations for the determination of the need for pressure regulators on center pivots and linear move systems.

Depth and Pressure Distributions

Sprinkler flow rates will vary as the lateral rotates on sloping land unless pressure regulators are used. Pressure regulators will be needed on each sprinkler head if the pressure varies by more than ±10% oven the length of the pivot at any point in the field. Pressure regulators are almost always required on low pressure systems on sloping land. Figure above shows general recommendations on whether or not pressure regulators are needed on a system. The selection of specific regulators depends on system pressures and the sprinkler selection. Flow control nozzles may also be an option but large fluctuations in pressures may adversely affect distribution patterns and droplet sizes.

Selection of the proper nozzles requires knowledge of the pressure distribution along the pivot lateral. This is complicated by the fact that the sprinkler discharges increase as you move toward the end of the pivot lateral while the pipe diameter remains constant (e.g., 6.38 inches [162 mm] diameter until the overhang after the last tower). The pressure, P_j, at point j in psi along the lateral is given by:

$$P_j = P_o - \frac{P_{lp} * L * f_p(R)}{100} - K E_g$$

where P_o is the pressure at the inlet to the pivot in psi, P_{lp} is the pressure loss in the pivot lateral pipe in psi per 100 ft, L is the radial length in feet to j, E_g is the elevation gain in feet at j, K is a units conversion factor of 0.4484 (use 0.1017 with meters and kPa) and f_p (R) is the dimensionless pressure distribution factor at distance R (at j). R is about 0.555 for most center pivots without an end gun and 0.56 with an end gun. A value of 0.36 is used for R on linear move systems due to the more equal distribution of flow from the outlets along the lateral. This can relationship also be calculated by the Hazen-Williams equation using a C factor of 140 or 145 for galvanized or epoxy lined steel pipe.

Distribution Patterns

Water application uniformity is an important performance criterion for the design and evaluation of center-pivot sprinkler irrigation systems. However, the water application depth of a center-pivot irrigation system is not uniform across a field. It depends on the sprinkler package, field topography, movement of the machine, and many other factors. Wind distortion of sprinkler distribution patterns is a major and dynamic factor.

Numerous coefficients of uniformity (CUs) have been developed over the past few decades. In general, all CUs can be divided into two categories: non-weighted and areal-weighted. Non-weighted CUs are calculated directly from the observations (actual or simulated catch-can data), and each observation is assumed to represent the same land area. Non-weighted CUs include Christiansen's CU_c, Wilcox and Swailes, Benami and Hore. Areal-weighted CUs are determined from both observations and the land area each observation represents. Areal-weighted CUs include CU_H by Heermann and Hein and the USDA-Soil Conservation Service pattern efficiency. For irrigation systems where each sprinkler covers the same amount of land area, non-weighted CUs can be used. On the other hand, if each sprinkler in an irrigation system covers different sized area, such as in a center-pivot irrigation system, areal-weighted CUs are preferred. The uniformity coefficient can range from 0.0 to 1.0, but the minimum desirable uniformity is about 0.85 for a center pivot irrigation systems. Both the application efficiency and uniformity coefficients are affected by the depth of irrigation.

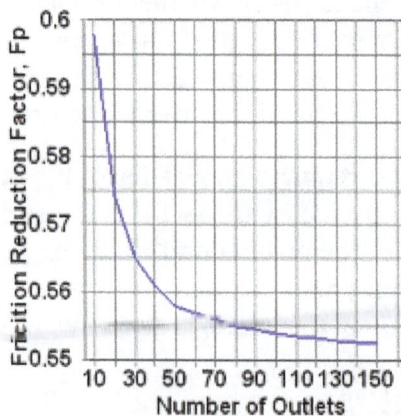

Friction reduction factors for multiple outlet pipelines.

Although a coefficient of uniformity can be used to compare different systems, it does not provide a functional description of actual applied water distribution. Therefore, statistical distribution functions are often used to represent the actual water application distribution. Many distribution functions have been proposed. Among them is the normal distribution, log normal, uniform and specialized power distributions.

One advantage of using distribution functions over a CU is that matching raw data to a theoretical function results in better estimation of the performance compared to the direct use of the raw data. When a distribution model is known, raw data are used to fit the distribution function and to obtain the function coefficients; and the coefficient of variance (CV) is calculated and used to compute the CU. The CV approach has been stated to be more appropriate than the CU approach in some cases. Distribution functions can also be used to determine the volume of over and under irrigation. This is done by simply integrating the product of the application depth and depth distribution function.

In using a distribution function to evaluate a sprinkler system design, two important points must be kept in mind. First, the distribution function is the probability distribution, not the actual spatial distribution, of water application depths. It can tell the probability of a given application depth, but it cannot tell which locations of the field actually receive the given amount of water. A distribution function may be adequate for evaluating the overall system performance, but it is not sufficient when a spatial water distribution is required as is needed, for example, in managing chemigation. Secondly, the distribution function is developed based on the assumption that water application depth is a random variable. That assumption may not be true if sprinkler distribution patterns have highly predictable shapes. In such a case, application depth may be directly calculated, and assuming a random distribution of application depth is vulnerable to errors and misinterpretations.

A center pivot irrigation model (CPIM) was developed to study the non-uniform distribution of irrigation water/nitrogen from the center pivot system. CPIM is a basic hydraulic model that considers topography and predicts nozzle pressures at any water emission point in the field. An empirical shape-pressure function specific to each sprinkler head is used to predict the water distribution from each head. CPIM overlaps the patterns and can produce maps showing the spatial distribution of water application depths. Complete field information on a center pivot system including actual topography and hydraulic data can be entered into the program's databases. There are also several "default" sprinkler (impacts only, at this time) packages and topographic options that can either be used directly or edited. CPIM is currently structured to fit within our defined GIS framework.

The application uniformity of the whole field can also be assessed. The CPIM results can be graphically compared with actual catch-can test results, if available. Results of the CPIM analysis could be used to target areas of highest potential nitrate leaching within a field for specific management practices.

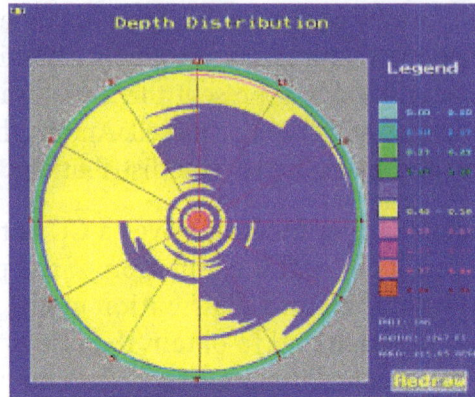

Example of water distribution from the CPIM model on
steeply sloping field without pressure regulators.

Chemigation

Center pivots provide an excellent vehicle to apply some chemicals and many fertilizers
to exactly match plant requirements. In some areas with very light soils as much as
80% of nitrogen fertilizer is applied through the center pivot system. Substantial crop
quality and pest control benefits may accrue when using this method.

The first rule of chemigation is safety. Special chemigation safety devices, check valves
and air relief valves are required for all chemical injection systems under federal and
state regulations. Well heads must be protected from reverse flows, system drainage
or back siphoning. Electric and hydraulic interlocks with time delays must be installed
between the injectors and irrigation pumps to prevent chemical injection when the irrigation system is not operating

Personnel must be specifically trained and, in many areas, licensed for chemical applications. Injection of any pesticide into an irrigation system must be specifically permitted by the pesticide label and may also be subjected to additional state regulations. Detailed records of all chemical applications need to be maintained for safety, evaluation,
legal and regulatory requirements.

All chemicals and chemical-water mixtures must be checked to avoid phytotoxic effects before any injection occurs. In addition, it is critical that all the chemicals being
injected at one time are compatible with each other and the water chemistry and concentration limits are not exceeded so that precipitates do not form. Emulsifiable pesticide concentrates and wetable powders may require special design and management
considerations (e.g., mechanical supply tank agitation) to help ensure uniform applications. Acidification to lower water pH may sometimes be required prior to injection
of the chemical.

Injection installations should always provide for complete mixing and uniform concentrations before the chemicals reach the field. Materials should be injected into the

center of the water flow to ensure quick dilution to reduce deterioration of piping, valving or other components. Generally, injection rates should not exceed 0.1% of the system water flow rate although concentration limits and label requirements for pesticides are usually less.

The use of positive displacement pumps is highly recommended for liquid chemical injections. The pumps should be adjustable and able to inject any water soluble chemical at low concentration levels (e.g., #100 ppm). The use of an in-line mixing chamber after injection may be necessary in some cases. A flow meter or other flow detection device can be connected to a controller that is programed to inject a specified amount of chemical from a nurse tank into the irrigation system at specific times based on flow rate.

Maintenance

Because center pivot and lateral move systems inherently supply small amounts of water on a frequent basis, soil water storage is usually limited. Consequently, an extended breakdown can be very serious in terms of yield reduction and expensive service calls.

Some system maintenance will always be required during the irrigation season. This typically involves replacing or repairing sprinkler heads or sprayers, leaky valves, flat tires, electrical shorts, oil leaks in gear boxes and breakdowns in gear boxes or drive line U-joint couplers.

Many in-season problems could be averted by a strong preseason maintenance program that includes: checking flanges, rubber flex boots, collector ring base drain and system drain valves for leaks; tightening nuts and bolts on flanges, trusses, tower supports, flex boot bands, lug nuts, etc.; greasing the pivot swivel, changing oil in gear boxes and replacing leaky gaskets as necessary; cleaning the pump panels and screens; cleaning or replacing filters, screens and ventilation/drain holes on engines, gear boxes and electrical panels; checking electrical systems including power cables, grounding conductors, power and pump shutdown wires, swivel connectors, pivot contactors; testing, replacing and repairing defective sprinkler heads, end gun bearings and pressure regulators; and checking that system operational water pressures are appropriate and pressure gauges are accurate.

Aggressive nozzle "management" and leak prevention programs can save water and energy. Nozzles become worn by silt and sand particles in the irrigation water leading to higher flow rates, less efficient pump operation and possible decreases in system pressures. Replacing worn nozzles on a regular basis, checking that the appropriate nozzles are installed in the correct locations and the installation of flow control nozzles/pressure regulators where needed will help ensure good uniformities of application and reduce overall water use.

Special Hardware Selection Considerations

Control Panels and Communications

There are a tremendous number of options available for control and operation of center pivots. Panels can be operated and adjusted manually or remotely by phone or radio. Rapid advances in computer and communications technologies have made it possible to remotely monitor as well as turn machines and injection pumps on and off from the farm office or even the front seat of the grower's pickup truck. As many as 100 or more machines may be controlled at one time. Alarms may sound when a machine stops for a multitude of reasons.

All the main manufactures are supplying advanced control panels at the pivot with remote communications capabilities. These are primarily digital devices with minimal electro-mechanical parts. Graphical interfaces can set system speed, run the system wet or dry, rate of application, system direction, end gun valves, fertilizer pumps, and other ancillary equipment. Different rotation speeds can be easily pre-programmed for different sections of a circle. The control panel will also keep a digital record of events for later analysis and record keeping.

These technologies are not inexpensive and the irrigator must determine the economic tradeoffs when selecting which panel and communication system (if any) will be purchased. Growers must evaluate their operation and maintenance costs and compare these with anticipated productivity gains and/or labor savings from various remote control options.

Tire Selection and Gear Boxes

Proper tire selection is critical to avoid problems with traction, deep ruts and easy crossing of the wheel tracks by farm equipment. Tires will vary with width and diameter ranging in size from about 24 to 38 inches (610 to 965 mm) in height and 11 to 17 inches (280 to 430 mm) in width. Generally, narrow larger diameter tires are used on heavy clay or loamy soils while wider tires are used on lighter sandy soils for greater flotation. Row crop farmers will often choose narrow tires where as growers of permanent crops such as alfalfa will normally select wide tires. Turf or tractor treads can be ordered although steel tires with heavy lugs or steel traction rims with heavy duty lugs mounted inside a regular rubber tire may be used on heavy clay soils where traction is a concern.

Obtaining good flotation can be a compromise of tire width and diameter. A narrow large diameter tire may have the same soil contact area or "foot print" as a wider small tower and provide equal flotation. Narrower tire tracks are usually easier to cross with equipment.

Gear boxes selection can be difficult. Worm gear systems are more expensive used on widely variable, steep terrain situations and on heavy soils with traction problems. Planetary and spur drive gear boxes are used in less demanding situations.

Screens and Sand Traps

When water is pumped directly from rivers, lakes or canals, the intakes should be equipped with self-cleaning screens. The stainless steel screens should be about twice the diameter of the attached pipeline and a mesh opening of about 0.25 inches (6 mm) or less. Cleaning is often accomplished with internal pressurized water jets that rotate inside the intake screen and push debris away from the mesh openings.

The pivot point should also have a stainless steel or galvanized screen with a mesh size of 0.1 inch (3 mm) or less to keep debris, algae, weed seeds, etc. from plugging nozzles. There needs to be a way to hydraulically isolate the screen from the rest of the system. These screens can be self-cleaning or manually cleaned. There should be a pressurized water supply for a hose to manually wash the screen in both cases.

Since sand and small gravel tend to collect at the distal end of pipelines, sand traps should also be placed at the distal end(s) of the center pivot or linear move system. These typically consist of a short section of 4-inch (100 mm) pipe pointing downward from a tee near the end of the mainline (near the end gun). These pipes have a 4-inch (100 mm) spring loaded valve or other method to quickly flush the collected sand from the system. Sometimes a special hose and large diameter (e.g., 0.25 in [6 mm] diameter) nozzle-spray plate arrangement is used to continuously flush the sand while the system is operating.

Garden Hose

A garden hose, hosepipe, or simply hose is a flexible tube used to convey water. There are a number of common attachments available for the end of the hose, such as sprayers and sprinklers (which are used to concentrate water at one point or to spread it over a large area). Hoses are usually attached to a hose spigot or tap.

Garden hoses are typically made of extruded synthetic rubber or soft plastic, often reinforced with an internal web of fibers. As a result of these materials, garden hoses are flexible and their smooth exterior facilitates pulling them past trees, posts and other obstacles. Garden hoses are also generally tough enough to survive scraping on rocks and being stepped on without damage or leaking.

Each male end of a typical garden hose can mate with the female connector on another, which allows multiple garden hoses to be linked end-to-end to increase their overall length. Small rubber or plastic washers (often confusingly called "hose washers") are used in female ends to prevent leakage, because the threads are not tapered and are not used to create a seal. Sometimes the gaskets stiffen, disintegrate, or fall out of older hoses, which results in pressurized leakage spraying from the hose; simply replacing the washer insert often fixes the problem.

Most garden hoses are not rated for use with hot water, and their packaging will often specify whether or not this is the case. Leaving non-reinforced hoses in the hot sun while pressurized can cause them to burst.

Hoses used to carry potable water are typically made of NSF International-listed polymers tested and shown not to leach harmful materials into the drinking water, such as the plasticizers (phthalates) used in polyvinyl chloride (PVC, or vinyl) hoses.

Usage

As implied by the name, garden hoses are commonly used to transport water for gardening, lawn care, and other landscaping purposes. They are also used for outdoor cleaning of items such as vehicles, equipment, building exteriors, and animals. NSF-approved hoses may be used for connecting drinkable water to recreational vehicles and trailers.

Sprayer pistol uses a quick-connect fitting, visible just beyond the sprayer grip

Whenever a flexible hose is connected to a drinkable water supply, the spigot or tap should be fitted with an approved backflow prevention device, to prevent contaminated water from being siphoned back, in the event of a pressure drop. Many water suppliers require this, and plumbing code may legally require permanently installed backflow preventers.

Porous or Perforated Hoses

Special hoses designed to leak throughout their length are sometimes used to gently distribute water on a lawn or garden. These hoses either have many small holes drilled or punched in them, or are deliberately formulated of a porous material, such as sintered rubber particles. These "soaker hoses" are a simple, low-cost, crude type of drip irrigation system.

Expandable Hoses

These differ from traditional hoses in that the inner membrane expands when filled with water, much like a balloon. An outer cover protects the delicate expandable membrane from punctures. Such hoses "grow" when pressurized, and shrink back down when the pressure is released, allowing for easier storage.

Standards and Connectors

Brass hose spigot with standard tapered pipe threads (internal, not visible) and GHT (visible) threads

Garden hoses connect using a male/female thread connection. The technical term for this arrangement is a "hose union". Spigots or sillcocks have male hose connectors only, and the mating end of a hose has a captive nut, which fits the threads there.

The thread standard for garden hose connectors in the United States, its territories, and Canada is known colloquially as "garden hose thread" (GHT), but its official designation is NH (NH stands for "National Hose"; 3⁄4-11.5NH is for full form threads as produced by cutting material such as the brass spigot outlet or hose male or female end fitting found on more expensive hoses; 3⁄4-11.5NHR is for thin-walled couplers produced by rolling thin material, usually brass, typically found on less expensive hoses; 3⁄4-14NPSH is for female hose ends that mate a hose to a tapered pipe thread without a spigot). The standard was defined by NFPA 1963, "Standard for Fire Hose Connections", then later by ANSI-ASME B1.20.7, which is $1 \frac{1}{16}$ inches (27 mm) diameter straight (non-tapered) thread with a pitch of 11.5 threads per inch (TPI). The female thread is abbreviated FHT, and the male part is abbreviated MHT. This fitting is used with $\frac{1}{2}$-inch, $\frac{5}{8}$-inch, and $\frac{3}{4}$-inch hoses.

In other countries, a British Standard Pipe (BSP) thread is used, which is $\frac{3}{4}$ inch and 14 TPI (male part outside diameter is 26.441 mm or 1.04 in). The GHT and BSP standards are not compatible, and attempting to connect a GHT hose to a BSP fitting, or vice versa, will damage the threads.

Various adaptors made of metal or plastic are available to interconnect GHT, BSP, NPT, hose barb, and quick connect fittings.

Quick Connectors

Starting in the 1990s, the use of quick-connector systems has become more popular. These are fittings that screw into the common hose connectors, allowing hoses and accessories to be easily connected together using a snap-fit system. The style manufactured by Gardena is common, imitated by and compatible with many other manufacturers. The connectors may optionally include an internal valve that is only opened by connecting the fitting, so that disconnecting a hose using this adaptor causes the water flow to stop. This greatly eases common tasks by allowing specialized sprayers to be interchanged without requiring two trips back to the spigot for each change.

Health Risks from Aerosols

In 2014, it was reported that use of common garden hoses in combination with spray nozzles may generate aerosols containing droplets smaller than 10 µm, which can be inhaled by nearby people. Water stagnating in a hose between uses, especially when warmed by the sun, can host the growth and interaction of *Legionella* and free-living amoebae (FLA) as biofilms on the inner surface of the hose. Clinical cases of Legionnaires' disease or Pontiac fever have been found to be associated with inhalation of garden hose aerosols containing *Legionella* bacteria. The report provided measured microbial densities resulting from controlled hose conditions in order to quantify the human health risks. The densities of *Legionella spp.* identified in two types of hoses were found to be similar to those reported during legionellosis outbreaks from other causes. It was proposed that the risk could be mitigated by draining hoses after use.

Garden hose in use

Closeup of a garden hose shows the crisscrossed fiber reinforcements

Gardena quick-connect hose fittings

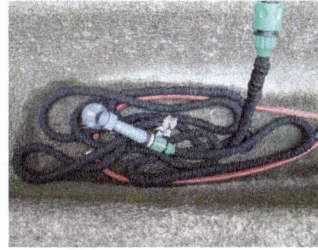
Special expandable hose can be
stretched to reach farther

Hose cart and some
quick-connect fittings

This greenhouse is equipped with
a hose suspended on pulleys

Subirrigation

Subirrigation is a greenhouse irrigation method that relies on capillary action to provide plants with water and nutrients from below their containers. The first documented subirrigation system was described in 1895, and several variations on the basic design were used for research purposes before the modern ebb-and-flow type systems emerged in 1974. Most subirrigation systems apply the fertilizer solution to a waterproof bench or greenhouse section, allowing the substrate to absorb the water through holes in the bottom of the containers. Because there is little or no leaching, subirrigation typically allows for the use of lower fertilizer solution concentrations. Although excess fertilizer salts typically accumulate in the top layer of the substrate, this does not seem to have a negative impact on plants. Subirrigation can conserve nutrients and water, reduce labor costs, and help growers meet environmental regulations. A challenge with subirrigation is the potential spread of pathogens via the fertilizer solution. When this is a concern, effective disinfection methods such as ultraviolet radiation, chlorine, or ozone should be used. Sensor-based irrigation control has recently been applied to subirrigation to further improve nutrient and water use efficiencies. Better control of irrigation may help reduce the spread of pathogens, while at the same time improving crop quality. The primary economic benefit of subirrigation is the reduction in labor costs, which is the greatest expenditure for many growers.

Greenhouse production is a vital part of the horticulture industry, with nearly 20,000 acres devoted to commercial greenhouse production in the United States alone. Frequent fertilization and irrigation is necessary, because plants typically grow in a small volume of substrate from which nutrients and water are rapidly depleted. Common greenhouse practices include overhead and drip irrigation, and plants are typically watered according to a set schedule. When more water is applied than can be absorbed by the substrate, excess water and nutrients leach. This leachate can be collected and reused, but often runs off to the external environment. Nutrients lost through runoff can contribute to nitrate and phosphate contamination of ground and surface water. Government regulations gradually require greenhouse growers to minimize their environmental impact by limiting nutrient runoff. Increasing fertilizer costs provide further incentive for growers to conserve nutrients and prevent losses.

Subirrigation is an irrigation technique that provides water or fertilizer solution to the bottom of containers. Capillary action of the substrate provides roots with water and nutrients. Greenhouse sections or benches with container-grown plants are periodically flooded within a closed system. The water or fertilizer solution is absorbed by the substrate through holes in the bottom of the container. The amount of water absorbed depends on substrate dryness, and irrigation volume closely matches plant requirements. Excess fertilizer solution is collected and reused for subsequent irrigation events, eliminating the environmental consequences associated with discharge, as well as reducing fertilizer costs for the producer. Subirrigation also reduces the cost of labor, which is the largest expenditure for horticultural producers.

Subirrigation has been used since at least 1895, when it was described by researchers at the Ohio Experimental Station. The subsequent development of hydroponics during the 1920s and sand-culture techniques in the 1930s helped establish the principles underlying modern subirrigation systems. Independently conceived subirrigation systems were also in use at the New Jersey and Purdue University agricultural experiment stations during the 1930s.

An early sand-culture system used two separate tanks to supply irrigation water and collected runoff from a sand bed. Excess water was eventually returned to the supply tank and electrical conductivity (EC) measurements were used to monitor nutrient concentrations of the recirculated solution. A later, more refined system provided uniform nutrient supply, increased aeration, and further minimized waste. In this nutrient-culture system, sealed benches containing an inert medium were periodically flooded with fertilizer solution, which was collected for reuse in the same reservoir used to flood the benches. Timers were eventually added to reduce labor costs. Other adaptations of these basic designs were developed and used for research and commercial production through the 1930s and 1940s. In all of these systems, plants grew directly in irrigated sand or other substrate, rather than in containers.

Section of sand-culture apparatus showing the sand bed, solution barrels, and plumbing; 1 inch (in.) = 2.54 cm, 1 ft (ft.) = 0.3048 m, and 50 mesh = 0.297 mm (0.0117 inch).

Subirrigation system for large-scale operation using two different substrates (porous media and fine gravel or cinders)

Subirrigation method of fertilizer solution culture.
At that time, all pipes were made by black iron

Subirrigation with containerized plants was first described in 1950 as an improved and simplified alternative to the sand-culture and nutrient-culture techniques. In these systems, relatively small sand beds in one compartment were subirrigated with fertilizer solution from a lower reservoir. This system was used to commercially produce african violet (*Saintpaulia ionantha*) and impatiens (*Impatiens walleriana*), and was later redesigned to accommodate larger plants for experimental purposes .Unlike modern subirrigation systems, sand was used as the growing medium and fertilizer solution had to be discharged periodically due to changes in nutrient concentration.

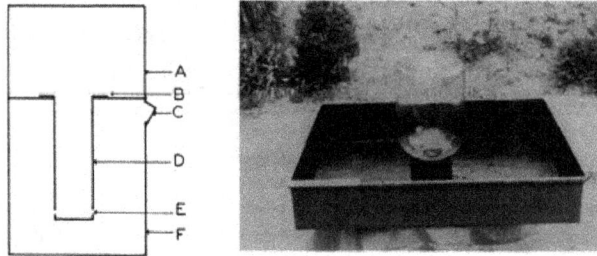

A double compartment container for subirrigation experiments in plant nutrition. Left: (A) upper compartment, (B) plastic collar, (C) opening for use in filling the lower compartment, (D) plastic tube, (E) perforation, and (F) lower compartment. Right: Sand-culture equipment coated with aluminum paint with the 1-gal (3.78 L) bottle reservoir. A 15-cm (5.9 inches) ruler on the lower left side indicates size

Twenty-two years later, a fully automated, highly versatile subirrigation system was developed. Individual containerized plants, grown in soilless substrate, were housed in an upper reservoir and subirrigated with fertilizer solution from a lower reservoir using a submersible pump activated by a timer. Unabsorbed fertilizer solution returned to the lower reservoir through a slow drain and could be reused indefinitely. The system drained more slowly than it was filled, thereby allowing sufficient time for uptake of fertilizer solution by capillary action, and maintaining substrate water content near container capacity throughout the growing period. This system could be easily modified for use with various container and reservoir sizes and configured so that one reservoir could supply multiple benches. Similar designs, now known as ebb-and-flow systems, subsequently became popular and are the most common type of subirrigation used in the greenhouse industry.

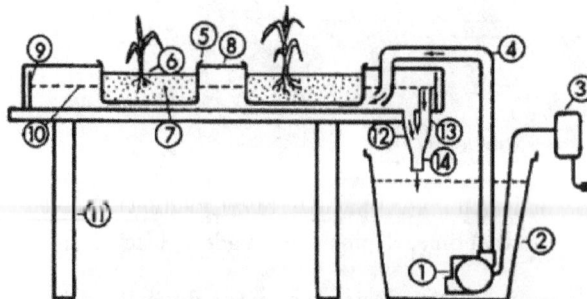

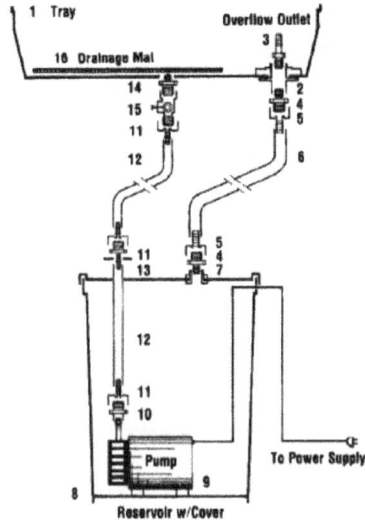

Pulsed subirrigation tray system

(1) structural foam tray (0.36 × 0.50 × 0.10 m), (2) bulkhead fitting (1/2-inch dual thread), (3) adapter [1/2-inch male pipe thread (MPT) × 1/2-inch barb], (4) adapter [1/2-inch MPT × male garden hose thread (MHT)], (5) swivel [female garden hose thread (FHT) × 1/2-inch barb], (6) garden hose (1/2-inch), (7) bulkhead fitting (1/2-inch single thread), (8) bucket (18 L) with lid, (9) submersible pump with 1/4-inch MPT outlet, (10) adapter (1/4-inch female pipe thread × MHT), (11) swivel (FHT × 1/4-inch barb), (12) polyethylene tubing (black 1/4-inch), (13) adapter (MHT × 1/4-inch barb), (14) adapter (MHT × 3/8-inch MPT), (15) ball valve (1/4-inch turn), (16) drainage mat; 1 m = 3.2808 ft, 1 inch = 2.54 cm. 1 L = 0.2642 gal.

Equipment

Subirrigation systems are classified as ebb-and-flow benches, flood-floor, trough-tray, wick system, mobile or Dutch trays, and capillary mat. A study evaluating subirrigation use in 50 greenhouse establishments in 26 states indicated that ebb-and-flow is used by 58% of the growers, flood-floor by 13%, and trough-tray systems by 8%, with 21% of the interviewees having two or more systems.

The typical ebb-and-flow system includes an elevated, water-tight bench where plants are grown, a fertilizer solution reservoir, and a pump .The bench is periodically flooded with fertilizer solution pumped from the reservoir to which any unabsorbed fertilizer solution eventually returns through a gravity drain at a sufficiently slow rate to allow for absorption. Irrigation frequency can be controlled using timers. Dutch trays, also known as mobile trays, are self-contained mobile ebb-and-flow benches that can easily be moved throughout a greenhouse. These trays are manufactured specifically for use in highly automated growing operations. While manufactured ebb-and-flow systems are common in commercial greenhouses, similar systems can be easily assembled from inexpensive materials to suit specific needs, and may be designed to operate without a pump. For example, used manual valves to flood children's swimming pools for production of large deciduous plants in containers.

The figure below shows types of equipment used in subirrigation.

Commercial prefabricated and automated ebb-and-flow benches for ornamental seedling and plant production	Small swimming pools adapted for large deciduous seedling production in containers

Flood-floor irrigation is a similar method in which plants are placed directly onto the greenhouse floor, typically made of concrete, and the entire space is flooded through holes in the floor. The floor has a gentle slope to allow the water to drain back to the fertilizer solution tank. This maximizes the proportion of greenhouse space used for production and operates more rapidly, but is often more labor-intensive than standard ebb-and-flow bench systems. A relatively new approach to flood-floor irrigation is the partial saturation ebb-and-flow watering, whereby the fertilizer solution is pumped onto the higher side of a sloped floor and flows down to drain at the low end. This creates a thin film of water on the floor and allows growers more control over how much water the substrate absorbs. Compared with standard flood-floors, partial saturation systems use less water and fertilizer and produce smaller plants.

In the trough-tray system, plants are placed in sloped shallow gutters or troughs. Fertilizer solution is pumped into the higher side of the troughs and runs down to a drain at the lower side and returns to the fertilizer solution tank. Trough-trays are most useful for operations that produce continuously in the same size containers. They are less versatile than ebb-and-flow benches because they can accommodate only a limited number of plants in relatively small containers, and take up more space for a similar number of plants compared with ebb-and-flow benches because the narrow troughs must be spaced apart. However, trough trays also allow for better air circulation through the canopy because of this spacing.

Other types of subirrigation include wick irrigation and capillary mats. Wick irrigation supplies water and nutrients from a fertilizer solution reservoir to the substrate via an absorptive wick, providing consistent moisture without runoff. Capillary mats are absorbent mats used to provide potted plants with moisture from below the containers to minimize fluctuations in substrate water content. A finely perforated, thin plastic film can be used to cover capillary mats to reduce evaporation and algal growth. Newer capillary mats may also have a liner below the mat, which prevents water from dripping from the mat. Capillary mats are versatile and can be used on a temporary basis and for outdoor production.

Principles Underlying Subirrigation, Substrates, and Containers

Water and nutrients are delivered to plants by the passive movement of water through the substrate due to capillary action. Substrate physical properties may affect the efficiency of capillary rise. Adequate media stability, density, particle structure, and water holding capacity are needed to allow water movement within the containers. Substrates with large particles have large pore spaces, reducing capillary action. Most commonly used soilless substrates are suitable for use with subirrigation, and different substrates can be mixed to suit particular needs.

Container height can also affect subirrigation efficiency because water must travel further to reach the upper part of the substrate in taller containers have shown that vertical gradients occur in subirrigated containers because water is absorbed by the lower part of the substrate and moves upward. Capillary rise can occur slowly; in 15-cm-diameter × 12-cm-tall pots filled with a peat–perlite substrate, water did not reach the upper substrate layer until up to 20 h after 'Panama Red' hibiscus (*Hibiscus acetosella*) plants were subirrigated.

While plastic containers are most common, containers made from biodegradable materials can also be used with subirrigation. compared the effects of using several types of biocontainers for growing cyclamen (*Cyclamen persicum*) using ebb-and-flow benches over a 15-week production cycle. Shoot dry weight was greater in all biocontainers than in the plastic control pots, with the exception of wood fiber containers. Plants grown in the wood fiber containers had reduced dry weights, presumably because these containers did not have holes in the bottom and water could not easily be absorbed through the walls. These containers also absorbed less nutrient solution per irrigation and had the shortest irrigation interval. Highly porous biocontainers (made from wood fiber, peat, dairy manure, and rice straw) required a higher total volume of nutrient solution over the course of the study because water readily evaporated from the container walls. These containers also had greatly reduced tensile strengths at the end of the production cycle. Bioplastic, rice hull, paper, and coconut fiber containers did not differ from the control containers in irrigation interval or total irrigation volume, and did not have reduced tensile strengths at the end of the production cycle. compared ebb-and-flow subirrigation to hand and drip watering using coleus (*Solenostemon scutellarioides*) grown in several types of biocontainers. Subirrigation improved growth in all treatments, and this effect was attributed to increased fertilization rate due to the absence of leaching. Subirrigation reduced the puncture strength of manure, paper, and wood fiber biocontainers, but not peat biocontainers.

Water Conservation

Subirrigation can be especially beneficial in areas where water scarcity is an issue. Compared with overhead irrigation, subirrigation systems have been consistently shown to reduce overall water use, primarily because excess water is collected and reused. For example, found that subirrigation requires 56% less water than overhead irrigation. Roeber indicated the different water use for potted plant production, showing that ebb-and-flow benches and troughs use 0.4 to 0.8 $m^3 \cdot m^{-2}$ water per year, drip irrigation use 0.8 to

1.6 m³·m⁻² water per year, and hand or sprinkler irrigation use 1.2 to 2.4 water m³·m⁻² per year. Subirrigation generally uses less water than overhead watering methods mainly because the unused solution is collected for reuse, rather than being lost due to drainage.

Salinity

In semiarid regions, growers often rely on saline water sources. Subirrigation can mitigate the effects of osmotic stress that can be induced by applying saline water. Tomato (*Solanum lycopersicum*) subirrigated with fertilizer solution prepared with saline water had fruit yields comparable to plants drip irrigated with the same solution. Another study demonstrated that subirrigated tomato yields were higher and salt accumulation was minimal when nutrient content of the irrigation solution was reduced by 30%. Subirrigation reduced tomato plant size and overall water use but improved water use efficiency. Accumulated salts were found primarily in the upper substrate layers. The accumulation of salts within the substrate prevented the fertilizer solution from becoming overly saline, reducing the need to discharge unusable fertilizer solution, resulting in more efficient water and fertilizer use. In zucchini squash (*Cucurbita pepo*), subirrigation with moderately saline water reduced growth rate and yield, but improved fruit quality and water use efficiency when compared with subirrigation with nonsaline nutrient solution.

Vegetable and Fruit/Tree Crops

Currently, subirrigation is primarily used for ornamental plant production. However, subirrigation may become an increasingly valuable method for the production of non-ornamental crops, such as vegetables and fruit/tree seedlings.

Subirrigation type	Species	Summary
Ebb-and-flow benches (1 × 1 × 0.15 m)	Cyclamen (*Cyclamen persicum*)	Plants grown in several types of biocontaners had higher shoot dry weights than those grown in plastic pots
Ebb-and-flow troughs (5 m²)	Chrysanthemum (*Dendranthema indicum*)	Used subirrigation to root chrysanthemum cuttings. Root growth was inversely correlated to substrate water content between waterings
Hand-watered plant saucers (0.195 m diameter)	Poinsettia (*Euphorbia pulcherrima*)	Subbirigation was compared with overhead watering. There was no difference in plant height or bract size, but subirrigated plants had lower dry weights and diameters
Ebb-and-flow benches (1.5 × 0.9 × 0.04 m)	'Panama Red' hibiscus (*Hibiscus acetosella*)	Subirrigaton was automated using soil moisture sensors
Potted plant system (dimensions not available)	Nephthylis (*Syngonoum podophyllym*)	Describes the design of a subirrigation system that operates without pumps
Hand-watered trays (0.5 × 0.35 × 0.06 m) lined with 6-mil black plastic	Rhododendron (*Rhododendron* sp.)	Hardwood stem cuttings had large root balls and a higher rooting percentage in lower pH media

Ebb-and-flow benches (2.4 × 1.2 m)	Petunia (*Petunia ×hybrida*) and wax begonia (*Begonia ×semperflorens-cultorum*)	Good growth and flowering over wide range of fertilizer solution concentrations, little effect of different substrates
Ebb-and-flow benches (2.4 × 1.2 m)	Petunia	Optimal fertilizer electrical conductivity (EC) decreases as temperature increases. Maintaining optimal leachate EC is better than maintaining constant fertilizer solution EC
Ebb-and-flow benches (2.4 × 1.2 m)	Alyssum (*Lobularia maritima*), celosia (*Celosia argentea*), dianthus (*Dianthus chinensis*), (*Gomphrena globosa*), stock (*Matthiola incana*), and zinnia (*Zinnia elegans*)	Optimal fertilizer concentrations were species-dependent. Superoptimal concentrations had little impact on growth
Ebb-and-flow benches (2.4 × 1.2 m)	Salvia (*Salvia splendens*)	High fertilizer concentrations increase the shoot:root ratio and hasten flowering
Ebb-and-flow benches (2.4 × 1.2 m)	Petunia and wax begonia	Optimal leachate EC is 1.0 to 1.7 dS·m^{-1}
Ebb-and-flow trays (1 × 1.1 m)	New guinea impatiens (*Impatiens walleriana*) and peace lily (*Spathiphyllum wallisii*)	Growth was most vigorous at 8 mM nitrogen (N) for new guinea impatiens and 10 mM N for peace lily, and declined thereafter. EC was two to five times higher in the upper substrate in all treatments
Ebb-and-flow benches (dimensions not available)	Areca palm (*Dypsis lutescens*) and philodendron (*Philodendron* sp.)	Compared with overhead watering, subirrigation used 10 times less water per unit area
Ebb-and-flow benches (dimensions not available)	Petunia	Subirrigated plants grew larger than overhead-watered plants at several rates of controlled release fertilizer incorporation. With liquid fertilizer, subirrigated plants were larger at 100 ppm N, but smaller at 150 ppm N
Flood floor (5 × 9 m) with a slope of 0.5%	Fortune spindle (*Euonymus fortunei*) and northern white-cedar (*Thuja occidentalis*)	Subirrigated plants had higher shoot dry weights at the lower of two fertilization rates. Both species grew less than overhead-watered plants in peat
Ebb-and-flow (dimensions not available)	Shamrock (*Oxalis regnellii* and *Oxalis triangularis*)	Overhead-watered plants were larger than subirrigated plants across a range of fertilization rates
Ebb-and-flow benches (1.5 × 1.8 m) and capillary mats	Geranium (*Pelargonium ×hortorum*)	Subirrigation was compared with microemitters and hand watering. Overall plant growth was similar in all treatments, but root numbers were higher in the subirrigated plants

Subirrigation for ornamental plant production

Ebb-and-flow benches (2.4 × 1.2 m)	Wax begonia	Optimal fertilizer concentration was estimated from water use efficiency
Ebb-and-flow benches (2.4 × 1.2 m)	Wax begonia and petunia	Light level did not affect optimal fertilizer solution concentration
Ebb-and-flow benches (2.4 × 1.2 m)	English ivy (*Hedera helix*)	Marketable plants were grown at the lowest fertilization rate (100 ppm N) in three light treatments

Ebb-and-flow benches (2.4 × 1.2 m)	Pale purple coneflower (*Echinacea pallida*)	Subirrigated plants had higher survival rates, greater height and biomass, and better N use efficiency than overhead-watered plants
Ebb-and-flow light weight 1.486 m² plastic benches	Croton (*Codiaeum variegatum*), dieffenbachia (*Dieffenbachia maculata*), and spathiphyllum (*Spathiphyllum ×petite*)	Plants grew larger in commercial potting mixes than in a custom peat:pine bark (1:1) mix. There was little response when N fertilization rate was increased above 4.2 g·L⁻¹
Trough-tray (5 × 0.16 m)	Geranium	Compared with a recirculating drip system, subirrigation led to higher substrate EC and less variation in nutrient solution (NS) EC. Subirrigated plants grew poorly at full strength NS
Ebb-and-flow (dimensions not available)	New guinea impatiens	Plant growth and quality decreased as substrate salinity increased. EC was highest in the upper substrate across treatments
Ebb-and-flow benches (0.9 × 1.5 m²)	Poinsettia	Post-production top watering can wash salts into root zone, but is not likely to cause damage
Capillary mat (1.8 × 1.8 m)	Pansy (*Viola ×wittrockiana*)	High fertilizer rates increase growth and shoot:root ratio
Capillary mat (1.8 × 1.8 m)	Marigold (*Tagetes patula*)	Drought stress reduces growth and plant height, but does not increase compactness
Hand-watered trays (0.51 × 0.13 × 0.06 m)	Chrysanthemum (*Dendranthema ×grandiflorum*), coleus (*Solenostemon scutellarioides*), red maple (*Acer rubrum*), and japanese tree lilac (*Syringa reticulata*)	Subirrigation can effectively replace misting for rooting cuttings of some ornamental crops

Subirrigation for vegetable production

Subirrigation type	Species	Summary
Hand-watered trays (0.51 × 0.13 × 0.06 m)	Lamarck serviceberry (*Amelanchier lamarckii*), amur maackia (*Maackia amurensis*), cherry (*Prunus serrulata*), spirea (*Spiraea ×bumalda*), lilac (*Syringa vulgaris*), and elm (*Ulmus glabra × Ulmus carpinijolia*)	Subirrigation without mist was effective for rooting softwood cuttings of elm, lilac, and spirea
Prototype subirrigation trays (0.62 × 0.41 × 0.15 m)	Rangpur lime (*Citrus limonia*)	Subirrigation induced higher plant growth and shortened the crop cycle compared with hand-water irrigation, anticipating the transplanting to the next phase (grafting)
Ebb-and-flow trays (1.22 × 1.22 m)	Northern red oak (*Quercus rubra*)	Under fertilized conditions, subirrigated seedlings had greater biomass and field diameter growth than top-watered plants
Subirrigation (dimensions not available)	Eucalyptus (*Eucalyptus grandis, Eucalyptus urophylla*, and *E. grandis × E. urophylla*)	Subirrigation increased the number of cuttings on clonal plants

Ebb-and-flow benches (2.44 × 1.22 × 0.115 m)	Koa (*Acacia koa*)	Subirrigated and overhead-watered seedlings had similar growth and vigor
Subirrigation system (3.75 × 0.85 m) with expanded clay	Coffee (*Coffea arabica*)	Adequate plant growth, regardless of the fertilization level of the stock plant
Subirrigation trays (0.62 × 0.41 × 0.15 m)	Rangpur lime	Validation of subirrigation technology for citrus rootstock production
Subirrigation system (dimensions not available)	Trifoliate orange (*Poncirus trifoliata*), citrange (*P. trifoliata* × *Citrus sinensis*), mandarin (*Citrus sunki*), and citrumelo (*P. trifoliata* × *Citrus paradisi*)	Different substrate composition influence on the capillary action
Subirrigation benches (1.65 × 2.55 × 0.2 m)	Araucaria (*Araucaria angustifolia*), jeriva (*Syagrus romanzoffiana*), cutieira (*Jownnesia princeps*), mutamba (*Guazuma ulmifolia*), red angico (*Anadenanthera macrocarpa*), and pink peroba (*Aspidosperma polyneuron*)	Subirrigation showed high water use efficiency and good plant uniformity in conic containers

Table: Selected publications related to subirrigation for fruit/tree seedling production.

Fertilization

Recirculation allows overall fertilizer use to be reduced because no nutrients are lost from the system. However, subirrigation requires careful management of fertilizer solution concentrations to produce high-quality greenhouse crops. Optimal fertilization rates for overhead irrigation systems are well known, but there is less applied information available about ideal fertilizer solution concentrations for subirrigation. Generally, fertilizer concentrations should be lower with subirrigation than with overhead or drip irrigation. Nutrient salts are not leached from the substrate and can accumulate within the containers, potentially exposing the plants to osmotic stress. High salinity occurs mainly in the upper substrate layer because salts move along with the capillary flow. As water evaporates from the surface, the salts accumulate in the top substrate layer. The accumulation of salts in the upper substrate layer is exacerbated by high fertilization rates. However, subirrigated plants are generally unaffected by high salinity in the upper substrate layers because root growth occurs primarily in the lower portions of the container where there is more water available. In the postproduction environment, accumulated salts from the upper portion of the substrate can potentially be washed to the bottom layers by top watering, but this does not seem to cause serious damage to the plants. Effective nutrient management for subirrigation requires minimizing the risk of osmotic stress, while providing the plants with adequate nutrition.

Optimal fertilizer solution concentrations vary among species and may depend on both the nutritional requirements and salt tolerance of a particular crop. This was demonstrated in a comparison of several bedding plant species subirrigated with various concentrations of Hoagland solution (12.5% to 200%). Full strength (100% concentration) Hoagland solution has a nitrogen concentration of 210 mg·L^{-1} and an EC of 2.0 dS·m^{-1}. Maximum zinnia (*Zinnia elegans*) and celosia (*Celosia argentea*) dry weights were observed at a 50% strength solution, and dry weights, as well as zinnia flower diameter, decreased at higher concentrations. In alyssum (*Lobularia maritima*) and dianthus (*Dianthus chinensis*), maximum growth occurred at the 100% concentration, but dianthus had the most flowers at a 200% concentration. Gomphrena (*Gomphrena globosa*) and stock (*Matthiola incana*) grew best within the 100% to 200% range, suggesting that these plants are particularly tolerant of accumulated salts . In a similar experiment, salvia (*Salvia splendens*) growth increased as fertilizer concentration increased from 12.5% to 100% Hoagland solution. Shoot dry weight, shoot:root ratio, and leaf area increased, while net photosynthesis, stomatal conductance, and transpiration decreased at higher fertilizer solution concentrations, suggesting that the treatment effects were due to shifts in carbon allocation and more efficient production of leaf area at increased fertilization rates. Leaf area decreased when fertilizer concentration was increased to 200%. Comparable effects of fertilizer solution concentration on plant growth were also observed in subirrigated pansy [*Viola×wittrockiana* (and wax begonia [*Begonia ×semperflorens-cultorum*]. However, found that fertilizer solution concentration had no effect on leaf area of subirrigated gerbera (*Gerbera jamesonii)* during the final production stage. Growth decreased when subirrigation fertilizer concentrations exceeded an optimum concentration in subirrigated poinsettia [*Euphorbia pulcherrima*], new guinea impatiens [*Impatiens ×hawkeri*], wax begonia , petunia [*Petunia ×hybrida*], miniature rose [*Rosa chinensis minima* ; collards (*Brassica oleraceavar.* acephala), kale (*B. oleracea* var. acephala), lettuce (*Lactuca sativa*), pepper (*Capsicum annuum*), and tomato transplants ; and shamrock species [*Oxalis regnellii* and *O. triangularis*].

Agrochemicals

Subirrigation allows pesticides and plant growth regulators (PGRs) to be applied to plants along with the fertilizer solution. Such applications should only be made if the label specifically allows for application by subirrigation. Using subirrigation to apply pesticides and PGRs ensures uniform application, prevents the release of these chemicals to the environment, and can further reduce labor costs as well as minimize employee exposure to pesticides. A potential problem with applying agrochemicals via subirrigation is that excess solution drains back into the holding tank. Contamination can be prevented by using separate holding tanks for agrochemicals, but this may be expensive. However, residual concentrations of PGRs in fertilizer solutions, such as paclobutrazol and uniconazole, were shown to be very low after application via subirrigation. Another consideration in applying agrochemicals through subirrigation is the substrate moisture level: drier substrate absorbs more solution, and thus agrochemical, than wetter substrates. Subirrigation has also been used for applying natural PGRs such as tea seed (*Camellia* sp.) powder.

Imidacloprid (Marathon 60 WP; OHP, Mainland, PA) applied via subirrigation to poinsettias was more effective at controlling silverleaf whiteflies (*Bemisia argentifolii*) than drench applications. Although drench applications resulted in faster initial uptake of the imidacloprid, subirrigation application resulted in higher leaf concentrations of imidacloprid after 63 d and better long-term control. Applications by subirrigation appeared to allow for plant uptake of imidacloprid over a longer period, resulting in more uniform distribution of the insecticide throughout the plant, as compared with drip irrigated plants treated with an imidacloprid drench.

Pathology

Transmission of root-infesting pathogens between containers or benches through recirculated irrigation water is a potential drawback of subirrigation. Oomycetes of the genera *Pythium* and *Phytophthora* are particularly problematic, since they produce large numbers of highly mobile aquatic zoospores and seriously affect plant growth and quality. Bacteria, viruses, and fungal pathogens can also migrate through the recirculating system and infect new hosts. Several measures can be taken to reduce the pathogen load of recirculated water. Infectious propagules can easily enter the reservoir with plant debris, and removal of dead material helps minimize disease spread. Filtering the fertilizer solution can also prevent contamination, removing extant microbes and preventing future infections. Membrane filters effectively remove pathogens, but are expensive and require frequent replacement. Slow media filtration is the passing of recirculated irrigation water through an inert media, most commonly sand, at a low flow rate. This process eliminates most plant pathogens through a combination of mechanical filtration and biological activity, but may be impractical because it requires a large amount of space, may not consistently provide adequate filtration, is relatively slow, and can support microbial populations that include human pathogens. Recirculated fertilizer solution can also be treated using ozone or ultraviolet radiation to eliminate pathogens. However, in a survey of greenhouses and nurseries, Meador et al. found that recirculating fertilizer solution from subirrigation systems, on average, did not meet recommended standards for horticultural water quality, and that microbial counts were exceptionally high. The spread of disease among containers in subirrigation systems is an important practical concern and more effective and economical preventative methods are needed.

Limiting the duration of flooding events or the amount of fertilizer solution provided to subirrigated plants may prevent or minimize disease spread. Partial saturation ebb-and-flow watering rapidly delivers water to plants. With this method, less water is absorbed by the substrate per flooding event. Elmer et al. demonstrated that partial saturation can prevent the spread of pythium root rot (*Pythium* sp.) infections. This may be because less fertilizer solution drains out of the containers following irrigation, fewer pathogens survive in the drier substrate, or partial saturation may provide plant root zone with more oxygen.

Sensor-based Subirrigation Control

Subirrigation is typically controlled using timers according to a predetermined schedule, usually designed to meet operational needs. Sensors can be used to monitor substrate moisture and control subirrigation based on plant water use. For a review of the use of different sensors in irrigation control, . Subirrigation has been successfully automated using tensiometers , lysimeters , and capacitance moisture sensors . Electronic switches connected to tensiometers to begin and end irrigation events at −5 and −1 kPa substrate matric potential (high and low media tension values, respectively) in drip and trough-tray irrigated containers were tested in several studies . Subirrigation with a standard nutrient solution decreased growth rate and yield of zucchini squash during the spring–summer growing season, but not during the summer–fall growing season compared with the automated drip system . When squash plants were irrigated with fertilizer solutions made with saline water, tensiometer-controlled subirrigation decreased yield, but improved fruit quality and increased water use efficiency . Another study used this automation with nonsaline water and showed that tensiometer-controlled subirrigation resulted in zucchini squash fruit yields equal to drip irrigation when full-strength fertilizer solution was used. However, use of half-strength fertilizer solutions reduced yield more with subirrigation than with drip irrigation. With zonal geranium (*Pelargonium×hortorum*) grown in the spring and fall at half- and full-strength fertilizer solutions, subirrigation only reduced growth when full-strength solution was used during the spring. In a similar comparison, found that drip irrigated plants used more water and were larger than plants grown in trough-trays when irrigation was automatically triggered at −7 kPa. However, when relatively low fertilizer concentrations were used, subirrigation resulted in equal yields as open-cycle drip irrigation. Triggering irrigation based on specific substrate moisture levels cannot only reduce water use; it can also be used to control plant vigor, potentially reducing the need for plant growth retardant applications. Inexpensive open-source microcontrollers can be used to build low-cost automated irrigation controllers.

Economic Benefits of Subirrigation

While fertilizer and water use can be reduced with subirrigation, this alone may not be sufficient to offset the initial price of installing an automated subirrigation system, which is often high. The primary economic benefit of subirrigation is that, by facilitating automation, it reduces the cost of labor, which is the greatest expenditure for many producers. Compared with traditional overhead watering, subirrigation may also improve plant health and quality. The incidence and spread of foliar disease is reduced because the leaves are not wetted during irrigation. Furthermore, subirrigated crops are typically uniform, because water and nutrients are evenly distributed, and uniformity makes handling and shipping easier.

References

- Bill Lauer (1 January 2004). AWWA Water Operator Field Guide. American Water Works Association. pp. 209–. ISBN 978-1-58321-315-5.

- Advantages-disadvantages-main-types-irrigation: lorecentral.org, Retrieved 12 May 2018

- Micro-irrigation: appropedia.org, Retrieved 13 April 2018

- Thomas, Jacqueline M.; Thomas, Torsten; Stuetz, Richard M.; Ashbolt, Nicholas J. (2014). "Your Garden Hose: A Potential Health Risk Due toLegionellaspp. Growth Facilitated by Free-Living Amoebae". Environmental Science & Technology. 48 (17): 10456–10464. doi:10.1021/es502652n. ISSN 0013-936X.

- Technical-manual-drip: godavaripipe.com, Retrieved 17 July 2018

- Spring-irrigation, agricultural-engineering: agritech.tnau.ac.in, Retrieved 26 June 2018

Chapter 3

Irrigation for Agriculture

Irrigation is a central feature in agriculture. It supports crop production, suppresses weed growth and prevents soil consolidation. A well-regulated irrigation practice is therefore of the utmost necessity in agriculture. This chapter closely examines some of the crucial aspects of irrigation in agriculture. It includes topics like agricultural water, irrigation scheduling, ditch irrigation, overhead irrigation, etc.

Agricultural Water

Agricultural water is water that is used to grow fresh produce and sustain livestock. The use of agricultural water makes it possible to grow fruits and vegetables and raise livestock, which is a main part of our diet. Agricultural water is used for irrigation, pesticide and fertilizer applications, crop cooling (for example, light irrigation), and frost control. According to the United States Geological Survey (USGS), water used for irrigation accounts for nearly 65 percent of the world's freshwater withdrawals excluding thermoelectric power. There are 330 million acres of land used for agricultural purposes in the United States that produce an abundance of food and other products.

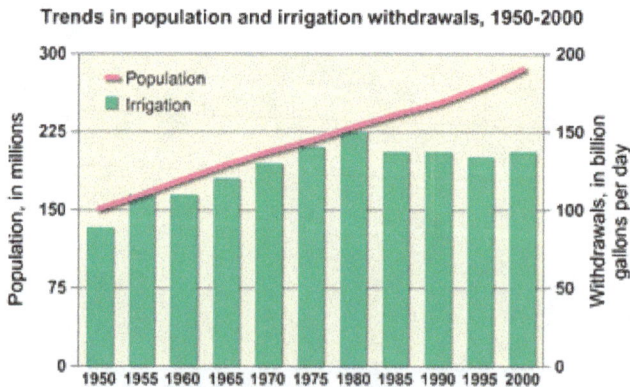

Trends in population and irrigation withdrawals, 1950-2000

When agricultural water is used effectively and safely, production and crop yield are positively affected. A decrease in applied water can cause production and yield to decrease. Management strategies are the most important way to improve agricultural water use and maintain optimal production and yield. The key is to implement management strategies that improve water use efficiency without decreasing yield. Some examples include improved irrigation scheduling and crop specific irrigation management. These strategies allow for the conservation of water and energy, and decrease grower's costs.

Sources of water used on the farm can be grouped into three types based on the likelihood that they can become contaminated:

(1) Surface water,

(2) Well water, and

(3) Municipal water.

Surface water includes ponds, open springs, lakes, rivers, and streams. It has the highest risk for contamination because we often do not have control over what might be entering the water source upstream at any given time. Access of wild and domestic animals, drainage from upstream cattle operations, runoff from manure piles, and sewage discharges are all possible causes for sudden and unexpected surface water contamination.

Water obtained from the wells on your farm generally has an intermediate risk. The potential for well water to become contaminated with harmful microorganisms is greatest when they are located too close to flood zones, septic tanks, cesspools, animal agricultural sites, manure storage areas, or drainage fields. Risks are greatly increased if the wells have not been constructed properly, or if the well casing has become cracked over time. However, if wells are properly sited, constructed, and maintained, they can be a reliable source of contaminant-free water.

Municipal water obtained from your local water authority has the lowest level of food safety risk. We expect this to be the safest type of water because it is required by law to meet the highest chemical and microbiological drinking water standards, and it is tested regularly to ensure that it is consistently safe to drink.

Preventing Agricultural Water from Becoming a Source of Contamination

Postharvest Water: Produce Washing, Handwashing, Cooling, Drinking

- Conduct a potable water test. It is critical to use pathogen-free water for all postharvest water used for washing, flumes and tanks, handwashing, and drinking.

- If postharvest water does not meet the drinking water standard, it may be possible to treat the source with one-time shock chlorination.

- If postharvest water does not meet the drinking water standard, it may be necessary to install a continuous sanitation system using chlorination or ultraviolet (UV) light.

Surface Water used for Irrigation

- Regularly monitor the microbial content of your surface water. Consider testing three times each season:

    o At planting

    o At peak use

    o At or near harvest

- Look for evidence of entry points for animals or areas where runoff can occur. Consider installing fences, vegetative buffer plantings, diversion berms, or other physical structures to protect the water from animal intrusion or drainage from contamination sources.

- When possible, use indirect irrigation methods, such as drip irrigation, that minimize water contact with fruits, tomatoes, peppers, and cole crops.

- Plasticulture methods that cover drip lines provide further protection for lower-growing crops such as leafy greens, cantaloupes, and summer squash.

- Use overhead irrigation in the morning to allow adequate drying of the crop surface before harvest. This speeds the destruction of both human and plant pathogens, and saves water.

- Maximize the time between overhead irrigation and harvest.

- Consider switching to well or municipal water for overhead irrigation and crop spraying methods.

Well Water

- Monitor your well water quality at least twice during the growing season.

- Check that your well is installed correctly. There should be at least 2 inches of grout maintained between the well casing and the surrounding soil to prevent infiltration of surface water. Because well drilling is not regulated in Pennsylvania, your well may not have the proper casing and grout to exclude surface water contaminants.

- Maintain a 100-foot radius around the well that is kept free from animal intrusion, manure piles, or other contamination sources.

- Install a sanitary well cap to prevent insects or small mammals from entering the well.

- Inspect your wells at least once each year. Check that the well cap and casing seal are in good condition.

Use of Water in Agriculture

Food and agriculture are the largest consumers of water, requiring one hundred times more than we use for personal needs. Up to 70 % of the water we take from rivers and groundwater goes into irrigation, about 10% is used in domestic applications and 20% in industry. Currently, about 3600 km3 of freshwater are withdrawn for human use. Of these, roughly half is really consumed as a result of evaporation, incorporation into crops and transpiration from crops. The other half recharges groundwater or surface flows or is lost in unproductive evaporation. Up to 90% of the water withdrawn for domestic use is returned to rivers and aquifers as wastewater and industries typically consume only about 5% of the water they withdraw. This wastewater from domestic sewage systems and industries should be treated before being dismissed.

Since the 1960s the global nutrition has considerably improved, providing more food per capita at progressively lower prices. This performance was possible through high-yielding seeds, irrigation and plant nutrition. As population keeps increasing more food and livestock feed need to be produced in the future and more water applied to this purpose. Irrigate agriculture will have to claim large quantities of water to produce the food required to feed the world. The main source of food for the population of the world is agriculture: this term also includes livestock husbandry, manages fisheries and forestry.

The composition of meals changes gradually as lifestyles change. What agriculture produces is driven by consumer demand, and changes in consumer preferences have an influence on the water needed for food production.

Cereals are by far the most important source of total food consumption: in developing countries the consumption of cereals 30 years ago represented 61% of total calories. It decreased to 56% nowadays and this reflects diet diversification, proving that more countries achieve higher levels of nutrition. It is expected that cereals will continue to supply more than 50% of the food consumed in the foreseeable future. A large proportion of cereals is produced for animal feed.

Food production from the livestock sector includes meat (beef, pork, poultry, etc.), dairy production and eggs.

For vegetative growth and development plants require water in adequate quantity and at the right time. Crops have very specific water requirements, and these vary depending on local climate conditions. The production of meat requires between six and twenty time more water than for cereals.

The following tables give an overview of the water consumption in food and agriculture.

Specific values for the water equivalent of a selection of food products are given in the first table. The second table shows the amount of water needed necessary for a few products per unit of consumption. A glass of wine acquires for example 120 liters of water, a hamburger 2.400 liters, a cotton shirt 4.000 liters and a couple of shoes made out of cows leather 8.000 liters.

The third table gives an overview of the amount of virtual water used in the different kind of agricultural products. The big difference between the countries is due to the climate, applied technology and the amount of production related to it.

Product	Unit	Equivalent water in m³ per unit
Cattle	Head	4000
Sheep	Head	500
Fresh beef	Kg	15
Fresh lamb	Kg	10
Fresh poultry	Kg	6
Cereals	Kg	1.5

Citrus fruits	Kg	1
Palm oil	Kg	2
Pulses, rood and tubers	Kg	1

Product	Quantity	Equivalent water in liters
Glass of beer	250 ml	75
Glass of wine	125 ml	120
Glass of milk	200 ml	200
Glass of apple juice	200 ml	190
Cup of coffee	125 ml	140
Glass of orange juice	200 ml	170
Cup of tea	250 ml	35
Chips bag	200 g	185
Slice of bread	30 g	40
Egg	40 g	135
Slice of bread with cheese	30 g+ 10 g	90
Hamburger	150 g	2400
Potato	100 g	25
Tomato	70 g	13
Apple	100 g	70
Orange	100 g	50
Cotton T- shirt	500 g	4100
Pair of shoes	1	8000
Sheet A4 paper	80 g/m²	10
Microchip	2g	32

Amount of virtual water per food per country in m³/ton

	U.S.	China	India	Russia	Indonesia	Australia	Brazil	Japan	Mexico	Italy	Netherlands	World average
Rice	1903	1972	4254	3584	3209	1525	4600	1822	3257	2506		3419
Wheat	849	690	1654	2375		1588	1616	734	1066	2421	619	1334
Corn	489	801	1937	1397	1285	744	1180	1493	1744	530	408	909
Soya beans	1869	2617	4124	3933	2030	2106	1076	2326	3177	1506		1789
Sugarcane	103	117	159		164	141	155	120	171			175
Cottonseed	2535	1419	8264		4453	1887	2777		2127			3644
Carton plaxel	5733	3210	18694		10072	4268	6281		4812			8242
Coconut		749	2255		2071		1590		1954			2545
Roast coffee	5790	7488	14500		21030		16633		33475			20682
Tea leaves		11110	7002	3002	9474		6592	4940				9205
Beef	13193	12560	16482	21028	14818	17112	16961	11019	37762	21167	11681	15497
Pork	3946	2211	4397	6947	3938	5909	4818	4962	6559	6377	3790	4856
Goat's meet	3082	3994	5l87	5290	4543	3839	4175	2560	10252	4180	2791	4043

Mutton	5977	5202	6692	7621	5956	6947	6267	3571	16878	7572	5298	6143
chicken	2389	3652	7736	5763	5549	2914	3913	2977	5013	2198	2222	3918
Eggs	1510	3550	7531	4919	5400	1844	3337	1884	4277	1389	1404	3340
Milk	695	1000	1369	1345	1143	915	1001	812	2382	861	641	990
Milk powder	3234	4648	6368	6253	5317	4255	4654	3774	11077	4005	2982	4602
Cheese	3457	4963	6793	6671	5675	4544	4969	4032	11805	4278	3190	4914
Cow leather	14190	13513	17710	22575	15929	18384	18222	11864	40482	22724	12572	16656

The amount of water involved in agriculture is significant and most of it is provided directly by rainfall. A rough calculation of global water needs for food production can be based on the specific water requirements to produce food for one person. The present average food ingest 2800 kcal/person/day may require 1000 m^3 per year to be produced. The world population is about 6 billion, so water needed to produce the necessary food, excluding water losses due to the irrigation system, is 6000 km^3. Most of it is provided by rainfall stored in the soil profile and only 15% is provided through irrigation. Irrigation therefore needs 900 km^3 of water per year for food crops. On average just about 40% of water withdrawn from rivers, lakes and aquifers for agriculture effectively contribute to crop production (the rest is lost through evaporation and deep infiltration). Consequently the current global water withdrawals for irrigation are estimated to be about 2000 to 2500 km^3 per year.

The irrigation level varies from area to area, mostly depending on climate conditions and on the development of irrigation infrastrure. The following figure shows the area equipped for irrigation as percentage of cultivated land by country.

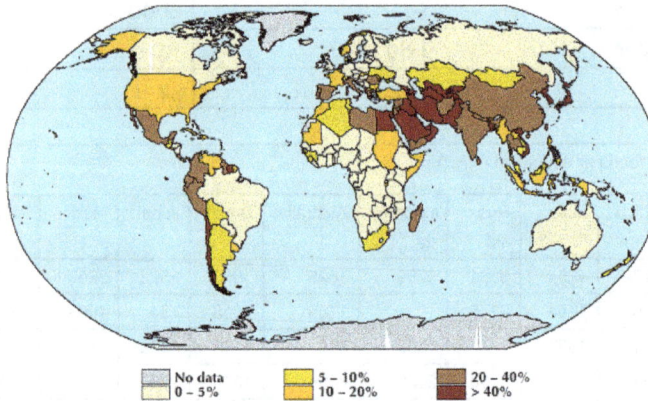

No data	5 – 10%	20 – 40%
0 – 5%	10 – 20%	> 40%

Irrigation-water management has a log way to adapt to the increasing production requirements, however water-saving technologies are already available and can significantly reduce the waste of water. If incentives are in place, as increasing the price of irrigation water, farmers will adopt water-saving irrigation technologies. The main technologies likely to be used in developing countries, where labor is normally abundant but capital scarce, are underground and drip irrigation. Both technologies depend on the frequent application of small amounts of water as directly as possible to the roots of crops. Reducing the pollution loads of water used by farms, industries and urban areas would enable much more of it to be re-used in irrigation. There are enormous potential benefits to be had from the use of wastewater for irrigation.

Agriculture will remain the dominant user of water at the global level. In many countries, in particular those situated in the arid and semi-arid regions of the world, this dependency can be expected to intensify. The contribution of irrigated agriculture to food production is substantial but in future the rate of growth will be lower than in the past. Both irrigated and non-irrigated agriculture still have scope for increasing productivity, including water productivity. Arguably, the expansion of irrigated agriculture protected people on the nutritional fringe from premature death, and preserved tracts of land under forest and wetlands from encroachment by hard-pressed farmers.

Role of Irrigation in Food Production

For vegetative growth and development, plants require, within reach of their roots, water of adequate quality, in appropriate quantity and at the right time. Most of the water a plant absorbs performs the function of raising dissolved nutrients from the soil to the aerial organs, from where it is released to the atmosphere by transpiration: agricultural water use is intrinsically consumptive. Crops have specific water requirements, and these vary depending on local climatic conditions. Whereas an indicative figure for producing one kilogram of wheat is about 1000 litres of water that is returned to the atmosphere, paddy rice may require twice this amount. The production of meat requires between six and twenty times more water than for cereals, depending on the feed/meat conversion factor. Specific values for the water equivalent of a selection of food products are given in table below. Water required for human food intake can be derived from these specific values in a grossly approximate way, depending on the size and composition of the meals.

Water requirement equivalent of main food products

Product	Unit	Equivalent water in m³ per unit	
Cattle	head	4000	
Sheep and goats	Head	500	This table gives examples of water required per unit of major food products.
Fresh beef	Kg	10	Including livestock, which consume the most water per unit.
Fresh lamb	Kg	10	Cereals, oil crops, and pulses, roots and tubers consume far less water.
Fresh poultry	Kg	6	
Cereals	Kg	1.5	
Citrus fruits	Kg	1	
Palm oil	Kg	2	
Pulses, root and tubers	Kg	1	

This table gives examples of water required per unit of major food products, including livestock, which consume the most water per unit. Cereals, oil crops, and pulses, roots and tubers consume far less water.

Food Production

Dominant Role of Rainfed Agriculture

Non-irrigated (rainfed) agriculture depends entirely on rainfall stored in the soil profile. This form of agriculture is possible only in regions where rainfall distribution ensures continuing availability of soil moisture during the critical growing periods for the crops. Non-irrigated agriculture accounts for some 60 percent of production in the developing countries. In rainfed agriculture, land management can have a significant impact on crop yields: proper land preparation leading surface runoff to infiltrate close to the roots improves the conservation of moisture in the soil. Various forms of rainwater harvesting can help to retain water in situ. Rainwater harvesting not only provides more water for the crop but can also add to groundwater recharge and help to reduce soil erosion. Other methods are based on collecting water from the local catchment and either relying on storage within the soil profile or else local storage behind bunds or ponds and other structures for use during dry periods. Recently, conservation agriculture practices such as conservation tillage have proven to be effective in improving soil moisture conservation.

The potential to improve non-irrigated yields is restricted where rainfall is subject to large seasonal and interannual variations. With a high risk of yield reductions or complete loss of crop from dry spells and droughts, farmers are reluctant to invest in inputs such as plant nutrients, high-yielding seeds and pest management. For resource-poor farmers in semi-arid regions, the overriding requirement is to harvest sufficient food stuff to ensure nutrition of the household through to the next harvest. This objective may be reached with robust, drought-resistant varieties associated with low yields. Genetic engineering has not yet delivered high-yield drought-resistant varieties, a difficult task to achieve because, for most crop plants, drought resistance is associated with low yields.

Role of Irrigation in Food Production

In irrigated agriculture, water taken up by crops is partly or totally provided through human intervention. Irrigation water is withdrawn from a water source (river, lake or aquifer) and led to the field through an appropriate conveyance infrastructure. To satisfy their water requirements, irrigated crops benefit from both more or less unreliable natural rainfall, and from irrigation water. Irrigation provides a powerful management tool against the vagaries of rainfall and makes it economically attractive to grow high-yield seed varieties and to apply adequate plant nutrition as well as pest control and other inputs, thus giving room for a boost in yields. Figure below illustrates the typical yield response of a cereal crop to water availability and the synergy between irrigation, crop variety and inputs. Irrigation is crucial to the world's food supplies. In 1998, irrigated land made up about one-fifth of the total arable area in developing countries but produced two-fifths of all crops and close to three-fifths of cereal production.

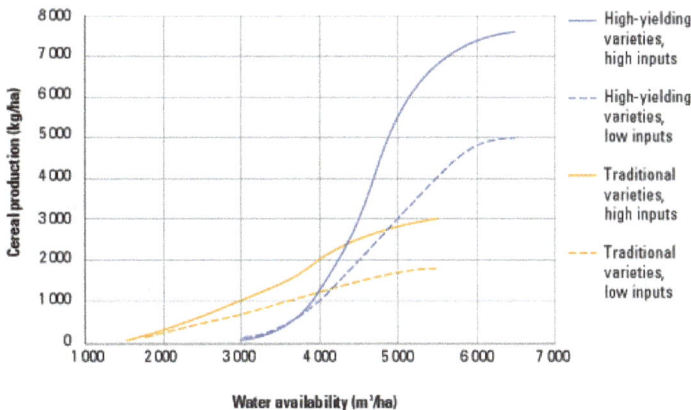

The developed countries account for a quarter of the world's irrigated area (67 million ha). Their annual growth of irrigated area reached a peak of 3 percent in the 1970s and dropped to only 0.2 percent in the 1990s. The population of this group of countries is growing only slowly and therefore a very slow growth in their demand and production of agricultural commodities is foreseen. The focus of irrigation development is consequently expected to be concentrated on the group of developing countries where demographic growth is strong. Increasing competition from the higher valued industrial and domestic sector results in a decrease in the amount of overall water allocated to irrigation. Figure below illustrates the case for the Zhang-he irrigation system in China.

Map below shows irrigated land as percentage of arable land in developing countries. A high proportion of irrigated land is usually found in countries and regions with an arid or semi-arid climate. However, low proportions of irrigated land in sub-Saharan Africa point also to underdeveloped irrigation infrastructure. Data and projections of irrigated land compared to irrigation potential in developing countries are shown in figure below. The irrigation potential figure already takes into account the availability of water. The graph shows that a sizeable part of irrigation potential is already used in the Near East/North Africa region (where water is the limiting factor) and in Asia (where land is often the limiting factor), whereas a large potential is still unused in sub-Saharan Africa and in Latin America.

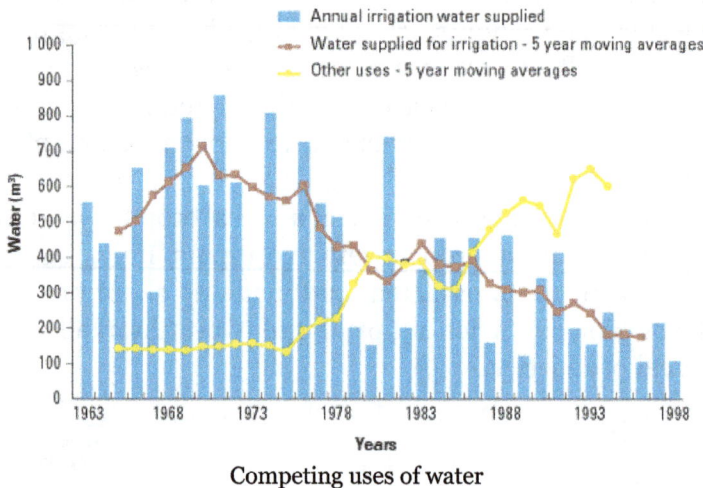

Competing uses of water

According to FAO forecasts, the share of irrigation in world crop production is expected to increase in the next decades. In particular in developing countries, the area equipped for irrigation is expected to have expanded by 20 percent (40 million ha) by 2030. This means that 20 percent of total land with irrigation potential but not yet equipped will be brought under irrigation, and that 60 percent of all land with irrigation potential (402 million ha) will be in use by 2030. The net increase in irrigated land (40 million ha, 0.6 percent per year) projected to 2030 is less than half the increase over the preceding 36 years (99 million ha, 1.9 percent per year). The projected slowdown in irrigation development reflects the projected lower growth rate of food demand, combined with the increasing scarcity of suitable areas for irrigation and of water resources in some countries, as well as the rising cost of irrigation investment. The first selection of economically attractive irrigation projects has already been implemented, and prices for agricultural commodities have not risen to encourage investment in a second selection of more expensive irrigation projects.

Most of the expansion in irrigated land is achieved by converting land in use in rainfed agriculture or land with rainfed production potential but not yet in use into irrigated land. The expansion of irrigation is projected to be strongest in South Asia, East Asia and Near East/North Africa. These regions have limited or no potential for expansion of non-irrigated agriculture. Arable land expansion will nevertheless remain an important factor in crop production growth in many countries in sub-Saharan Africa, Latin America and some countries in East Asia, although to a much smaller extent than in the past. The growth in wheat and rice production in the developing countries will increasingly come from gains in yield, while expansion of harvested land will continue to be a major contributor to the growth in production of maize.

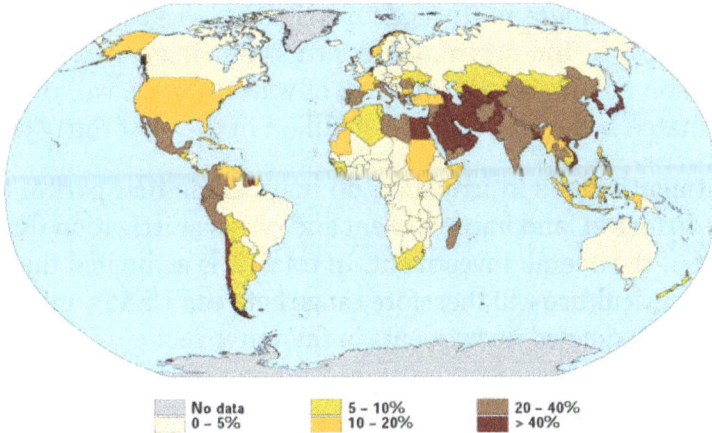

No data	5 – 10%	20 – 40%
0 – 5%	10 – 20%	> 40%

Area equipped for irrigation as percentage of cultivated land by country

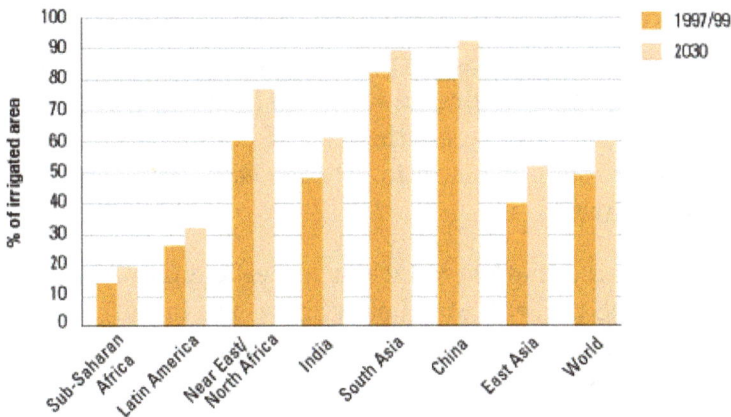

Irrigated area as proportion of irrigation potential in developing countries

Future Investments in Irrigation

In many developing countries, investments in irrigated infrastructures have represented a significant share of the overall agricultural budget during the second half of the twentieth century. The unit cost of irrigation development varies with countries and types of irrigated infrastructures, ranging typically from US$1 000 to US$10 000 per hectare, with extreme cases reaching US$25 000 per hectare (these costs do not include the cost of water storage as the cost of dam construction varies on a case-by-case basis). The lowest investment costs in irrigation are in Asia, which has the bulk of irrigation and where scale economies are possible. The most expensive irrigation schemes are found in sub-Saharan Africa, where irrigation systems are usually smaller and developing land and water resources is costly.

In the future, the estimates of expansion in land under irrigation will represent an annual investment of about US$5 billion, but most investment in irrigation, between US$10 and 12 billion per year, will certainly come from the needed rehabilitation and modernization of aging irrigated schemes built during the years 1960-1980. In the 1990s, annual investment in storage for irrigation was estimated at about US$12 billion (WCD, 2000). In the future, the contrasting effects of reduced demand for irrigation expansion and increased unit cost of water storage will result in an annual investment estimated between US$4 and 7 billion in the next thirty years.

Typically, investment figures in irrigation do not include that part of the investment provided by the farmer in land improvement and on-farm irrigation that can represent up to 50 percent of the overall investment. In total, it is estimated that annual investment in irrigated agriculture will therefore range between US$25 and 30 billion, about 15 percent of annual expected investments in the water sector.

Water use Efficiency

Assessing the impact of irrigation on available water resources requires an estimate of total abstraction for the purpose of irrigation from rivers, lakes and aquifers. The volume extracted is considerably greater than the consumptive use for irrigation because of conveyance losses from the withdrawal site to the plant root zone. Water use efficiency is an indicator often used to express the level of performance of irrigation systems from the source to the crop: it is the ratio between estimated plant requirements and the actual water withdrawal.

On average, it is estimated that overall water use efficiency of irrigation in developing countries is about 38 percent. Map below shows the importance of agriculture in the countries' water balance, and figure above shows the expected growth in water abstraction for irrigation from 1998 to 2030. The predictions are based on assumptions about possible improvements in irrigation efficiency in each region. These assumptions take into account that, from the farmer's perspective, wherever water is abundant and its cost low, the incentives to save water are limited. Conversely, if farmers can profitably irrigate more land using their allocation in an optimum way, irrigation efficiency may reach higher levels.

Improving irrigation efficiency is a slow and difficult process that depends in large part on the local water scarcity situation. It may be expensive and requires willingness, know-how and action at various levels. Table below shows current and expected water use efficiency for developing countries in 1998 and 2030, as estimated by FAO. The investment and management decisions leading to higher irrigation efficiency are taken and involve irrigation system management and the system-dependent farmers. National water policy may encourage water savings in water-scarce areas by providing incentives and effectively enforcing penalties. When upstream managers cannot ensure conveyance efficiency, there may be no incentives for water users to make efficiency gains. With groundwater, this caveat may not apply since the incentive is generally internalized by the users, and in many cases groundwater users show much greater efficiency than those depending on surface resources. Provides an overview of different aspects of potential improvements in agricultural water use efficiency.

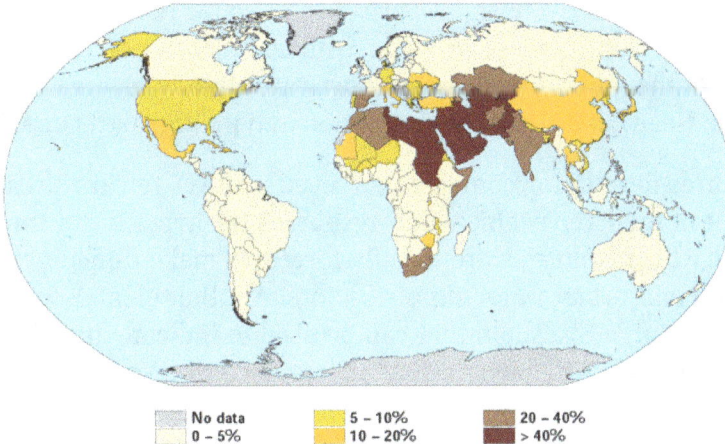

Agricultural water withdrawals as percentage of renewable water resources

Future Water Withdrawals for Irrigation

Irrigation water withdrawal in developing countries is expected to grow by about 14 percent from the current 2 130 km3 per year to 2 420 km3 in 2030. This finding is based specifically on individual assessments for each developing country. Harvested irrigated area (the cumulated area of all crops during a year) is expected to increase by 33 percent from 257 million ha in 1998 to 341 million ha in 2030. The disproportionate increase in harvested area is explained by expected improvements in irrigation efficiency, which will result in a reduction in gross irrigation water abstraction per ha of crop. A small part of the reduction is due to changes in cropping patterns in China, where consumer preference is causing a shift from rice to wheat production.

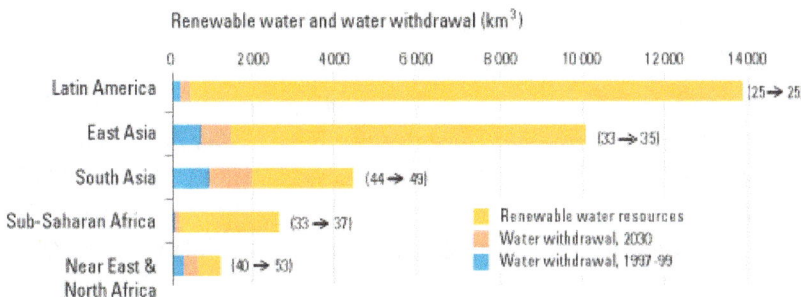

Irrigation and water resources: current and predicted withdrawals

	Sub-Saharan Africa	Latin America	Near East & North Africa	South Asia	East Asia	All countries
Water use efficiency in irrigation (%)						
1998	33	25	40	44	33	38
2030	37	25	53	49	35	42
Irrigation water withdrawal as a percentage of renewable water resources (%)						
1998	2	1	53	36	8	8
2030	3	2	58	41	8	9

Water use efficiency in 1998 and 2030 (predicted) in 93 developing countries

While some countries have reached extreme levels of water use for agriculture, irrigation still represents a relatively small part of total water resources of the developing countries. The projected increase in water withdrawal will not significantly alter the overall picture. At the local level, however, there are already severe water shortages, in particular in the Near East / North Africa region and in large parts of Asia.

Of the ninety-three developing countries surveyed by FAO, ten are already using more than 40 percent of their renewable water resources for irrigation, a threshold used to flag the level at which countries are usually forced to make difficult choices between their agricultural and urban water supply sectors. Another eight countries were using more than 20 percent, a threshold that can be used to indicate impending water scarcity. By 2030 South Asia will have reached the 40 percent level, and the Near East and North Africa not less than 58 percent. However, the proportion of renewable water resources allocated to irrigation in sub-Saharan Africa, Latin America and East Asia in 2030 is likely to remain far below the critical threshold.

Special Role of Groundwater

Water contained in shallow underground aquifers has played a significant role in developing and diversifying agricultural production. This is understandable from a resource management perspective: when groundwater is accessible it offers a primary buffer against the vagaries of climate and surface water delivery. But its advantages are also quite subtle. Access to groundwater can occasion a large degree of distributive equity, and for many farmers, groundwater has proved to be a perfect delivery system. Because groundwater is on demand and just-in-time, farmers have sometimes made private investments in groundwater technology as a substitute for unreliable or inequitable surface irrigation services. In many senses, groundwater has been used by farmers to break out of conventional command and control irrigation administration. Some of the management challenges posed by large surface irrigation schemes are avoided, but the aggregate impact of a large number of individual users can be damaging, and moderating the 'race to the pump-house' has proved difficult. However, as groundwater pumping involves a direct cost to the farmer, the incentives to use groundwater efficiently are high. These incentives do not apply so effectively where energy costs are subsidized; such distortion has arguably accelerated groundwater depletion in parts of India and Pakistan.

The technical principles involved in sustainable groundwater and aquifer management are well known but practical implementation of groundwater management has encountered serious difficulties. This is largely due to groundwater's traditional legal status as part of land property and the competing interests of farmers withdrawing water from common-property aquifers. Abstraction can result in water levels declining beyond the economic reach of pumping technology; this may penalize poorer farmers and result in areas being taken out of agricultural production. When near the sea, or in proximity to saline groundwater, over-pumped aquifers are prone to saline intrusion. Groundwater quality is also threatened by the application of fertilizers, herbicides and pesticides that percolate into aquifers. These 'non-point' sources of pollution from agricultural activity often take time to become apparent, but their effects can be long-lasting, particularly in the case of persistent organic pollutants.

Fossil groundwater, that is, groundwater contained in aquifers that are not actively recharged, represent a valuable but exhaustible resource. Thus, for example, the large sedimentary aquifers of North Africa and the Middle East, decoupled from contemporary recharge, have already been exploited for large-scale agricultural development in a process of planned depletion. The degree to which further abstractions occur will be limited in some cases by the economic limits to pumping, and promoted where strong economic demand from agriculture or urban water supply becomes effective. Two countries, Libyan Arab Jamahiriya and Saudi Arabia, are already using considerably more water for irrigation than their annual renewable resources, by drawing on fossil groundwater reserves. Several other countries rely to a limited extent on fossil groundwater for irrigation. Where such groundwater reserves have a high strategic value in terms of water security, the depletion of such reserves to irrigate is questionable.

Role of Irrigation in Alleviating Poverty and Improving Food Security

There is a positive, albeit complex, link between water services for irrigation and other farm use, poverty alleviation and food security. Many of the rural poor work directly in agriculture, as smallholders, farm laborers or herders. The overall impact can be remarkable: in India, for example, in unirrigated districts 69 percent of people are poor, while in irrigated districts, only 26 percent are poor. Their income can be boosted by pro-poor measures, such as ensuring fair access to land, water and other assets and inputs, and to services, including education and health. Relevant reforms of agricultural policy and practices can strengthen these measures.

The availability of water confers opportunities to individuals and communities to boost food production, both in quantity and diversity, to satisfy their own needs and also to generate income from surpluses. Irrigation has a land-augmenting effect and can therefore mean the difference between extreme poverty and the satisfaction of the household's basic needs. It is generally recognized that in order to have an impact on food security, irrigation projects need to be integrated with an entire range of complementary measures, ranging from credit, marketing and agricultural extension advice to improvement of communications, health and education infrastructure. Land tenure may also represent a major constraint: irrigation schemes controlled by absentee landowners and serving distant markets, even when highly efficient, may fail to improve local food security when both commodities and benefits are exported.

Irrigation projects are as diverse as the local situations in which they are implemented. Generally, small-scale irrigation projects including projects based on shallow groundwater pumping provide a manageable framework that can give control to the local poor and avoid leaking resources to the non-poor. Large-scale irrigation, as may be determined by the need to carry out large-scale engineering works to harness water and convey it to the fields, can also be made to work for the poor provided that the benefits can be shared equitably, and investment, operation and maintenance costs are efficiently covered.

Managing Agricultural Risk for Sustainable Livelihoods

Community-managed small-scale irrigation systems, by improving yields and cropping intensities, have proved effective in alleviating rural poverty and eradicating food insecurity. Marketing of agricultural produce, both locally and at more distant places when adequate transport and communication infrastructure is available, can make a significant contribution to the income of farmers. Bank deposits and credits, as well as crop insurance, can be used to finance agricultural operations and buffer against climatic risk. However, banking services usually are not accessible to people who have no collateral assets. Many rural credit systems may also not accommodate pay-back over a period of years – the time needed to realize the benefits from investment in irrigation technology. However, non-conventional credit systems based on trust and social solidarity can support poor farmers. Improvement in poor people's food commodity storage facilities reduces post-harvest losses and can save significant amounts of food, thus contributing to food security. Similarly, where technically and financially possible, storing water in surface reservoirs and aquifers is a strategic means for managing agricultural risk. Water in the reservoir or the aquifer is, in a way, equivalent to money in the bank.

Irrigation Contributes to Creating Off-farm Employment

Irrigation, supported through inputs such as high-yielding varieties, nutrients and pest management, together with a more extended agricultural season, higher cropping intensity and a more diverse assortment of crops, can generate rural employment in other non-agricultural services. The productivity boost provided by irrigated agriculture results in increased and sustained rural employment, thereby reducing the hardships experienced by rural populations that might otherwise drift to urban areas under economic pressure. Growth in the incomes of farmers and farm laborers creates increased demand for basic non-farm products and services in rural areas. These goods and services are often difficult to trade over long distances. They tend to be produced and provided locally, usually with labour-intensive methods, and so have great potential to create employment and alleviate poverty. Studies in many countries have shown multipliers ranging from two (in Malaysia, India and the United States) to six.

The Contribution of Fisheries and Aquaculture to Food Security

Fish has a very good nutrient profile and is an excellent source of high-quality animal protein and of highly digestible energy. Refugees and displaced people facing food insecurity may turn to fishing for survival, where this possibility exists. Staple-food fish, often low-valued fish species, is in high demand in most developing countries, because of its affordability.

Inland fish production provides significant contributions to animal protein supplies in many rural areas. In some regions freshwater fish represent an essential, often irreplaceable source of cheap high-quality animal protein, crucial to a well-balanced diet in marginally food-secure communities. Most inland fish produce is consumed locally, marketed domestically, and often contributes to the subsistence and livelihood of poor people. The degree of participation in fishing and fish farming is high in many rural communities. Fish production is often undertaken in addition to agricultural or other activities. Yields from inland capture fisheries, especially subsistence fisheries, can be very significant, even though they are often greatly under-reported. Yields from inland capture fisheries are highest in Asia in terms of total volumes, but are also important in sub-Saharan Africa. Fishery enhancement techniques, especially stocking of natural and artificial water bodies, are making a major contribution to the total catch.

Rural aquaculture contributes to the alleviation of poverty directly through small-scale household farming of aquatic organisms for domestic consumption and income. It also contributes indirectly through employment of the poor as service providers to aquaculture or as workers on aquatic farms. Poor rural and urban consumers can greatly benefit from low-cost fish provided by aquaculture. For aquaculture to be effective in alleviating poverty, it should focus on low-cost products favored by the poor, and emphasis should be placed on aquatic species feeding low in the food chain. The potential exists for aquaculture production for local markets and consumers. Combined rice-fish systems are possible and carry high benefits because they provide cereals and protein at the same time. These systems have also been shown to have beneficial effects on the malaria situation, where mosquito vectors breed in rice fields and where the selected fish species feed on the mosquito larvae.

Forestry and Food Security

A large number of forest products contribute to food security: FAO estimates that about 1.6 billion people in the world rely to a certain extent on forest resources for their livelihoods. For most rural people in the world, firewood provides the fuel for cooking food, and so its availability is an integral part of food security. The bio-energy sector generates employment and income in developing countries.

Most forests and tree plantations subsist on rainwater, or may develop around irrigation schemes. Some tree species can use large amounts of water drawn from water storage in the soil profile and in shallow aquifers. Trees in forests and outside of forests provide significant benefits to poor people and contribute to food security. The benefits from water used by forests are largely seen in terms of wood and non-wood products, as well as in protecting the environment, reducing soil degradation, and preserving biodiversity.

Irrigation Scheduling

Scheduling irrigation is attempting to apply water to potatoes at the appropriate amount for a specific stage in the plants development and growth. The potato plant's use of water is primarily for transpiration and tuber production and, therefore, irrigation is most important from emergence to vine senescence. Transpiration is the movement of water through the vine, from roots to leaves, to compensate for water loss at stomates (pores) that are open to allow gas exchange (oxygen and carbon dioxide) between the plant and the atmosphere thereby supporting plant photosynthesis and respiration.

Principles of Irrigation Scheduling from Soil-water View

1. Soil moisture. Sites for monitoring soil moisture should be chosen to be most representative of the field. The purpose is to limit under-watering of the heavier soils and over-watering lighter soils. For precision irrigation where watering can be controlled in smaller areas within the field, more monitors would be needed and both better and poorer soils would need to be monitored.

2. Root zone depth. Root zone depth is the zone where most of the root structure is found. This varies with different potato varieties but as a general rule, roots develop down to 18 inches below the seed piece.

3. Water holding parameters. Two measurements would be important. The "fill point" is the wettest a soil can be before water drains below the root zone. This

would be near 100% field capacity (FC) or 100% holding capacity of the root zone and depends on soil texture. In general, sandy soils have the lowest FC while silt loams have the highest with clays being intermediate. The "refill point" is the driest a soil can be before daily water use is lowered due to too little water in the root zone. This begins to induce the shutting of stomates resulting in reduction of carbohydrate synthesis (photosynthesis) and respiration (metabolism), and leads to wilting. This has a direct relation to yield. The difference between field capacity and 40% depreciation is the "allowable depletion" (AD) amount of water and, for potato, is 20-25% FC or about half the total available water (about 40% FC). In sandy loam soils, the AD is three-quarters to one-inch water up to a depth of 12 inches or one to one and a half inch for the root zone of a full-grown determinate potato variety. Soils that are compacted or tend to seal will lower water-holding capacity and reduce penetration of water into the soil.

4. Effective irrigation. Effective irrigation is the amount of water that actually gets into the root zone and is available to the plant. Some of the irrigation water (actual irrigation) is lost as run-off, evaporation or deep percolation.

5. Daily water use. Daily water usage by the potato is dependent on the growth stage of the plant and environmental conditions on that day. It is directly related to canopy development, mostly leaves which contain nearly all the stomates. Environmental conditions that affect daily water use are air temperature, relative humidity, wind, and solar radiation. An excellent guide to daily water usage is evapotranspiration (ET) data that is calculated from weather station data. Most, if not all, land-grant universities like University of Nebraska calculate ETs for stations around the state and provide them on the web and for publication in local newspapers. ET is the total daily water use from both transpiration by the plant and evaporation from the soil and can be as high as a third of an inch in a day for potato. ET needs also to be adjusted for canopy size or row closure, canopy width divided by row spacing; 80% row closure is considered full canopy or 100% ET. Therefore, a potato crop that has row-closed on sandy loam at field capacity (three-quarters to one inch AD) can be carried for two and a half to three days under high ET conditions. From this, one can estimate a two to two and a half inches weekly irrigation requirement for potatoes during tuber bulking under high ET conditionshigh temperature, low relative humidity, intense sunlight, high winds, and long days. Seasonal ETs differ for crops due to duration at full canopy and growing season.

Key Factors in Managing Irrigation

- How much water gets into the soil,

- How much water the soil can hold, and

- How much water is being used by the plant.

The accurate determination of an irrigation schedule is a time-consuming and complicated process. The introduction of computer programs, however, has made it easier and it is possible to schedule the irrigation water supply exactly according to the water needs of the crops. Ideally, at the beginning of the growing season, the amount of water given per irrigation application, also called the irrigation depth, is small and given frequently. This is due to the low evapotranspiration of the young plants and their shallow root depth. During the mid season, the irrigation depth should be larger and given less frequently due to high evapotranspiration and maximum root depth. Thus, ideally, the irrigation depth and/or the irrigation interval vary with the crop development

When sprinkler and drip irrigation methods are used, it may be possible and practical to vary both the irrigation depth and interval during the growing season. With these methods it is just a matter of turning on the tap longer/shorter or less/more frequently.

When surface irrigation methods are used, however, it is not very practical to vary the irrigation depth and frequency too much. With, in particular, surface irrigation, variations in irrigation depth are only possible within limits. It is also very confusing for the farmers to change the schedule all the time. Therefore, it is often sufficient to estimate or roughly calculate the irrigation schedule and to fix the most suitable depth and interval; in other words, to keep the irrigation depth and the interval constant over the growing season. Three simple methods to determine the irrigation schedule are briefly described: plant observation method, estimation method and simple calculation method.

The plant observation method is the method which is normally used by farmers in the field to estimate "when" to irrigate. The method is based on observing changes in plant characteristics, such as changes in colour of the plants, curling of the leaves and ultimately plant wilting.

The simple calculation method is based on the estimated depth (in mm) of the irrigation application, and the calculated irrigation water need of the crop during the growing season.

Plant Observation Method

The plant observation method determines "when" the plants have to be irrigated and is based on observing changes in the plant characteristics, such as changes in colour of the plants, curling of the leaves and ultimately plant wilting. The changes can often only be detected by looking at the crop as a whole rather than at the individual plants. When the crop comes under water stress the appearance changes from vigorous growth (many young leaves which are light green) to slow or even no growth (fewer young leaves, darker in colour, and sometimes greyish and dull).

Some crops react to water stress by changing their leaf orientation: with adequate water available, the leaves are perpendicular to the sun. However, when little water is available, the leaves turn away from the sun (thus reducing the transpiration and production).

To use the plant observation method successfully, experience is required as well as a good knowledge of the local circumstances. A farmer will, for example, know where the sandy spots in the field are, which is where the plants will first show stress characteristics: the colour changes and wilting are more pronounced on the sandy spots.

The disadvantage of the plant observation method is that by the time the symptoms are evident, the irrigation water has already been withheld too long for most crops and yield losses are already inevitable. It is important to note that it is not advisable to wait for the symptoms. Especially in the early stages of crop growth, irrigation water has to be applied before the symptoms are evident

Another indicator of water availability is the leaf temperature. If the leaves are cool during the hot part of the day, the plants do not suffer from water stress. However, if the leaves are warm, irrigation is needed. Special devices have been developed to measure the leaf temperature in relation to the air temperature. However, they must be calibrated for specific conditions before being used to determine the irrigation schedule.

Another method used to determine the irrigation schedule involves soil moisture measurements in the field. When the soil moisture content has dropped to a certain critical level, irrigation water is applied. Instruments to measure the soil moisture include gypsum blocks, tensiometers and neutron probes. Their use, however, is beyond the scope of this manual.

Estimation Method

Estimating the Irrigation Schedule

Here, a table is provided to estimate the irrigation schedule for the major field crops during the period of peak water demand; the schedules are given for three different soil types and three different climates. The table is based on calculated crop water needs and an estimated root depth for each of the crops under consideration. The table assumes that with the irrigation method used the maximum possible net application depth is 70 mm.

With respect to soil types, a distinction has been made between sand, loam, and clay, which have, respectively, a low, a medium and high available water content. With respect to climate, a distinction is made between three different climates.

Shallow and/ or sandy soil	In a sandy soil or a shallow soil (with a hard pan or impermeable layer close to the soil surface), little water can be stored; irrigation will thus have to take place frequently but little water is given per application
Loamy soil	In a loamy soil more water can be stored than in a sandy or shallow soil. Irrigation water is applied less frequently and more water is given per application.
Clayey soil	In a clayey soil even more water can be stored than in a medium soil. Irrigation water is applied even less frequently and again more water is given per application.
Climate 1	Represents a situation where the reference crop evapotranspiration ETo = 4 - 5 mm/day.
Climate 2	Represents an ETo = 6 - 7 mm/day.
Climate 3	Represents an ETo = 8 - 9 mm/day.

An overview indicating in which climatic zones these ETo values can be found is given below:

Climatic zone	Mean daily temperature		
	low (less than 15°C)	medium (15-25°C)	high (more than 25°C)
Desert/arid	4 - 6	7 - 8	9 - 10
Semi-arid	4 - 5	6 - 7	8 - 9
Sub-humid	3 - 4	5 - 6	7 - 8
Humid	1 - 2	3 - 4	5 - 6

It is important to note that the irrigation schedules are based on the crop water needs in the peak period. It is further assumed Chat little or no rainfall occurs during the growing season. Some examples on the use of Table are given below:

1. Estimate the irrigation schedule for groundnuts grown on a deep, clayey soil, in a hoc and dry climate.

Firstly, the climatic class has to be identified: climate 3 (ETo = 8-9 mm/day) represents a hot climate. Table above shows that for climate 3 the interval for groundnuts grown on a clayey soil is 6 days and the net irrigation depth is 50 mm. This means that every 6 days the groundnuts should receive a net irrigation application of 50 mm.

2. Estimate the irrigation schedule for spinach grown on a loamy soil, in an area with an average temperature of 12° C during the growing season.

The average temperature is low: climate 1 (ETo = 4-5 mm/day). Table above shows, with climate 1, for spinach, grown on a loamy soil an interval of 4 days and a net irrigation depth of 20 mm.

3. Estimate the irrigation schedule of sorghum grown on a sandy soil, in an area with a temperature range of 15-25° C during the growing season.

The average temperature is medium: climate 2 (ETo = 6-7 mm/day). Table above shows, with climate 2 for sorghum grown on a sandy soil, an irrigation interval of 6 days and a net irrigation depth of 40 mm.

Adjusting the Irrigation Schedule

a. Adjustments for the non-peak periods

The irrigation schedule, which is obtained using table above, is valid for the peak period; in other words, for the mid-season stage of the crop.

During the early growth stages, when the plants are small, the crop water need is less than during the mid-season stage. Therefore, it may be possible to irrigate during the early stages of crop growth, with the same frequency as during the mid-season, but with smaller irrigation applications. It is risky to give the same irrigation application as during the mid-season, but less frequently; the young plants may suffer from water shortage as their roots are not able to take up water from the lower layers of the root zone.

Dry harvested crops or crops which are allowed to die before harvest (for example grain maize) need less water during the late season stage than during the mid-season stage (the peak period). During the late season stage, the roots of the crops are fully developed and therefore the same amount of water can be stored in the root zone as during the mid-season stage. It is thus possible to irrigate during the late season stage less frequently but with the same irrigation depth as during the peak period.

In summary, in order to save water, it may be feasible to irrigate, during the early stages of the crop development, with smaller irrigation applications than during the peak period. During the late season stage it may be feasible to irrigate less frequently, in particular if the crop is harvested dry.

When adjusting the irrigation schedule for the non-peak periods, it should always be kept in mind that the irrigation schedules must be simple, in particular in surface irrigation schemes where many farmers are involved. It will often be necessary to discuss with the farmers, before implementing the irrigation schedule, the various alternatives and come to an agreement which best satisfies all parties involved.

b. Adjustment for climates with considerable rainfall daring the growing season

The schedules obtained from table above are based on the assumption that little or no rainfall occurs during the growing season. If the contribution from the rainfall is considerable during the growing season, the schedules need to be adjusted: usually by making the interval longer. It may also be possible to reduce the net irrigation depth. It is difficult to estimate to which values the interval and the irrigation depth should be adjusted. It is therefore suggested to use the simple calculation method, instead of the estimation method, in the case of significant rainfall during the growing season. Alternatively it is possible to adjust the irrigation schedule to the actual rainfall.

c. Adjustment for local irrigation practices or irrigation method used

It may happen that the net irrigation depth obtained from table above is not suitable for the local conditions. It may not be possible, for example, to infiltrate 70 mm with the irrigation method used locally. Tests may have shown that it is only possible to infiltrate some 50 mm per application.

In such cases, both the net irrigation depth and the interval must be adjusted simultaneously. For example, suppose that maize is grown on a clayey soil in a moderately warm climate. According to table above, the Interval is 10 days and the net irrigation depth is 70 mm. This corresponds to an irrigation water need of 70/10 = 7 mm/day.

Instead of giving 70 mm every 10 days, it is also possible to give:

63 mm every 9 days

56 mm every 8 days

49 mm every 7 days

42 mm every 6 days etc.

This means that in the above example an interval of seven days is chosen with a net application depth of 49 mm.

d. Adjustment for shallow soils

A soil which is shallow can only store a little water, even if the soil is clayey. For shallow soils - sandy, loamy or clayey - the column "shallow and/or sandy soil" of Table should be used.

e. Adjustment for salt-affected soils

In the case of irrigating salt-affected soils, special attention needs to be given to the determination of the irrigation schedule.

Application of the Simple Calculation Method

The simple calculation method to determine the irrigation schedule is based on the estimated depth (in mm) of the irrigation applications, and the calculated irrigation water need of the crop over the growing season.

Unlike the estimation method, the simple calculation method is based on calculated irrigation water needs. Thus, the influence of the climate, i.e. temperature and rainfall, is more accurately taken into account. The result of the simple calculation method will therefore be more accurate than the result of the estimation method.

The simple calculation method to determine the irrigation schedule involves the following steps that are explained in detail below:

Step 1:	Estimate the net and gross irrigation depth (d) in mm.
Step 2:	Calculate the irrigation water need (IN) in mm, over the total growing season.
Step 3:	Calculate the number of irrigation applications over the total growing season.
Step 4:	Calculate the irrigation interval in days.

Step 1: Estimate the net and gross irrigation depth (d) in mm

The net irrigation depth is best determined locally by checking how much water is given per irrigation application with the local irrigation method and practice. If no local data are easily available, table above can be used to estimate the net irrigation depth (d net), in mm. As can be seen from the table, the net irrigation depth is assumed to depend only on the root depth of the crop and on the soil type. It must be noted that the d net values in the table are approximate values only. Also the root depth is best determined locally. If no data are available, table below can be used which gives an indication of the root depth of the major field crops.

Approximate net irrigation depths, in mm

	Shallow rooting crops	Medium rooting crops	Deep rooting crops
Shallow and/or sandy soil	15	30	40
Loamy soil	20	40	60
Clayey soil	30	50	70

Approximate root depth of the major field crops

Shallow rooting crops (30-60 cm):	Crucifers (cabbage, cauliflower, etc.), celery, lettuce, onions, pineapple, potatoes, spinach, other vegetables except beets, carrots, cucumber.
Medium rooting crops (50-100 cm):	Bananas, beans, beets, carrots, clover, cacao, cucumber, groundnuts, palm trees, peas, pepper, sisal, soybeans, sugarbeet, sunflower, tobacco, tomatoes.
Deep rooting crops (90-150 cm):	Alfalfa, barley, citrus, cotton, dates, deciduous orchards, flax, grapes, maize, melons, oats, olives, safflower, sorghum, sugarcane, sweet potatoes, wheat.

Not all water, which is applied to the field, can indeed be used by the plants. Part of the water is lost through deep percolation and runoff. To reflect this water loss, the field application efficiency (ea) is used. The gross irrigation depth (d gross), in mm, takes into account the water loss during the irrigation application and is determined using the following formula:

$$d\,grass = \frac{100 \cdot d\,net}{ea}$$

d gross = gross irrigation depth in mm

d net = net irrigation depth in mm

ea = field application efficiency in percent

If reliable local data are available on the field application efficiency, these should be used. If such data are not available, the following values for the field application efficiency can be used:

- for surface irrigation	: ea = 60%
- for sprinkler irrigation	: ea = 75%
- for drip irrigation	: ea = 90%

If, for example, tomatoes are grown on a loamy soil, Tables 4 and 5 show that the estimated net irrigation depth is 40 mm. If furrow irrigation is used, the field application efficiency is 60% and the gross irrigation depth is determined as follows:

$$d\,grass = \frac{100 \cdot 40}{60} = 67\,mm = rounded\,65\,mm$$

Step:2 Calculate the irrigation water need (IN) in - over the total growing season

Assume that the irrigation water need (in mm/month) for tomatoes, planted 1 February and harvested 30 June, is as follows:

	Feb.	Mar.	Apr.	May	June
IN (mm/month)	IN (mm/month)	IN (mm/month)	IN (mm/month)	IN (mm/month)	IN (mm/month)

The irrigation water need of tomatoes for the total growing season (Feb–June) is thus (67 + 110 + 166 + 195 + 180 =) 718 mm. This means that over the total growing season a net water layer of 718 mm has to be brought onto the field.

If no data on irrigation water needs are available, the estimation method should be used.

Step 3: Calculate the number of irrigation applications over the total growing season

The number of irrigation applications over the total growing season can be obtained by dividing the irrigation water need over the growing season (Step 2) by the net irrigation depth per application (Step 1).

If the net depth of each irrigation application is 40 mm (d net = 40 mm; Step 1), and the irrigation water need over the growing season is 718 mm (Step 2), then a total of (718/40 =) 18 applications are required.

Step 4: Calculate the irrigation interval (INT) in days

Thus a total of 18 applications is required. The total growing season for tomatoes is 5 months (Feb–June) or 5 x 30 – 150 days. Eighteen applications in 150 days corresponds to one application every 150/18 = 8.3 days.

In other words, the interval between two irrigation applications is 8 days. To be on the safe side, the interval is always rounded off to the lower whole figure: for example 7.6 days becomes 7 days; 3.2 days becomes 3 days.

In this example, the irrigation schedule for tomatoes is as follows:

d net = 40 mm

d gross = 65 mm

interval = 8 days

Adjusting the Simple Calculation Method for the Peak Period

When using the simple calculation method to determine the irrigation schedule, it is advisable to ensure that the crop does not suffer from undue water shortage in the months of peak irrigation water need.

For instance, in the above example the interval is 8 days, while the net irrigation depth is 40 mm. Thus every 30 days (or each month): 30/8 x 40 mm = 150 mm water is applied. The amount of water given during each month (d net) should be compared with the amount of irrigation water needed during that month (IN).

The result is shown below. The "IN" values represents the irrigation water needs, while the "d net" values represent the amount of water applied. The "d net - IN" values show whether too much or too little water has been applied:

	Feb	Mar	Apr	May	June	Total
IN (mm/month)	87	110	166	195	180	718
d net (mm/month)	150	150	150	150	150	750
d net - IN (mm/month)	+83	+40	-16	-45	-30	+32

The total net amount of irrigation water applied (750 mm) is more than sufficient to cover the total irrigation water need (718 mm). However, in February and March too much water has been applied, while in April, May and June, too little water has been applied.

Care should be taken with under-irrigation (too little irrigation) in the peak period as this period normally coincides with the growth stages of the crops that are most sensitive to water shortages.

To overcome the risk of water shortages in the peak months, it is possible to refine the simple calculation method by looking only at the months of peak irrigation water need and basing the determination of the interval on the peak period only.

In the example given above for tomatoes, this means looking at the months April, May and June:

Months of peak irrigation water need	Apr	May	June	Sub-total
IN (mm/month)	166	195	180	541

The total irrigation water need from April to June (90 days) is 541 mm, while the net irrigation depth is 40 mm. Thus 541/40 = 13.5 (rounded 14) applications are needed. Fourteen applications in 90 days means one application every 6.4 (rounded 6) days. Calculated this way the irrigation schedule for the tomatoes would be:

d net = 40 mm

d gross = 65 mm

interval = 6 days

Over the total growing period of 150 days, this means 150/6 = 25 applications, each 40 mm net and thus in total 25 x 40 = 1000 mm.

The overall result of adjusting the irrigation schedule to the months of peak irrigation water demand is shown below:

	Feb	Mar	Apr	May	June	Total
IN (mm/month)	67	110	166	195	180	718
d net (mm/month)	200	200	200	200	200	1000
d net - IN (mm/month)	+133	+90	+34	+5	+20	+282

This way of determining the irrigation schedule avoids water shortages in the month of peak water needs but on the other hand also results in a higher seasonal irrigation water application.

It is possible to combine the two schedules. In this way some water is saved, and there are no water shortages in the peak period, but it is a bit more complicated for the farmers.

The result of the combined irrigation schedule for the whole growing season is as follows:

	Feb	Mar	Apr	May	June	Total
IN (mm/month)	67	110	166	195	180	718
d net (mm/month)	150	150	200	200	200	900
d net - IN (mm/month)	+83	+90	+40	+5	+20	+182

In summary:

Feb-March

d net = 40 mm

d gross = 65 mm

Interval = 8 days

April-May-June

d net = 40 mm

d gross = 65 mm

Interval = 6 days

Adjusting the Irrigation Schedule to Actual Rainfall

The estimation method to determine the irrigation schedule can only be used when no significant rainfall occurs during the growing season. The simple calculation method is based on the average irrigation water need of the crop which is the average crop water need minus the average effective rainfall. This method is used when designing and implementing an irrigation system with a "rotational" water supply: each field receives a certain amount of water on dates that are already fixed in advance. The rotational supply takes into account the average rainfall only and thus does not take into account the actual rainfall; this results in over-irrigation in wetter than average years and under-irrigation in drier than average years. In surface irrigation systems the rotational water supply method is most commonly used.

There are also water supply methods which allow the irrigation water to be distributed "on demand". The farmer can take water whenever necessary. In this case it is possible to take the actual rainfall into account and thus give the correct amount of irrigation water even in drier or wetter years. With this method of irrigation scheduling, however, the rainfall has to be measured on a daily basis. The net irrigation depth (d net) has to be determined in accordance with the irrigation method used. In addition, the crop water need has to be known on a daily basis for each month of the growing season. As soon as the accumulated water deficit exceeds the value of the net irrigation depth, irrigation water is supplied.

An example is given below for a situation with a crop water need (CWN) of 8 am/day and a net irrigation depth (d net) of 45 mm. As soon as the accumulated deficit exceeds the d net (= 45 mm), irrigation water is supplied. Note that the "deficit" can never be positive; maximum zero.

day	CWN (mm/day)	Rain (mm)	d net (mm)		Accumulated deficit (mm)
1	8	-	-		-8
2	8	-	-	(-8-8)	-16
3	8	-	-	(-16-8)	-24
4	8	-	-	(-24-8)	-32
5	8	-	-	(-32-8)	-40
6	8	-	45	(-40-8+45)	-3
7	8	-	-	(-3-8)	-11

8	8	12	-	(-11-8+12)	-7
9	8	24	-	(-7-8+24)	0
10	8	-	-	(0-8)	-8
11	8	-	-	(-8-8)	-16
12	8	-	-	(-16-8)	-24
13	8	4	-	(-24-8+4)	-28
14	8	-	-	(-28-8)	-36
15	8	-	-	(-36-8)	-44
16	8	-	45	(-44-8+45)	-7
17	8	-	-	(-7-8)	-15
etc.					

In the above example of adjusting the irrigation schedule to the actual rainfall, irrigation takes place on day 6, on day 16, etc. with on each occasion a net irrigation depth of 45 mm.

Advantages of Irrigation Scheduling

1. It enables the farmer to schedule water rotation among the various fields to\ minimize crop water stress and maximize yields.

2. It reduces the farmer's cost of water and labor through fewer irrigations, thereby making maximum use of soil moisture storage.

3. It lowers fertilizer costs by holding surface runoff and deep percolation (leaching) to a minimum.

4. It increases net returns by increasing crop yields and crop quality.

5. It minimizes water-logging problems by reducing the drainage requirements.

6. It assists in controlling root zone salinity problems through controlled leaching.

7. It results in additional returns by using the "saved" water to irrigate non-cash crops that otherwise would not be irrigated during water-short periods.

Ditch Irrigation

Irrigation ditches are manmade channels that deliver water to homes, farms, industries and other human uses. Most ditches divert water from natural creeks and rivers and bring it to other areas.

Ditches have headgates on the creeks that they divert. Most headgates are operated manually by ditch company personnel. Ditch headgates can be closed, when needed, to prevent water from being diverted from the creek into the ditch.

Ditches are typically owned, operated, and maintained by private companies. Water in ditches is allocated by shares issued by the company. Those shares represent proportional amounts of water.

Only those who have rights or shares in the ditch are allowed to remove water from it

Ditch companies are responsible for keeping the ditch channels clear of debris or sediment that would obstruct the flow of water. Barrow (roadside) ditches are the maintenance responsibility of the adjacent property owner – including culverts that run under driveways.

Overhead Irrigation

Overhead sprinkler systems spray water into the air above and around the foliage of the crop in a broadcast pattern. The water droplets should be large enough to fall through the canopy to growing substrate or to run along the foliage and stems to the base of the crop.

Sizes of droplets are largely determined by the sprinkler system to achieve distribution. Overhead sprinklers are 70 to 75 percent efficient in water delivery; evaporation occurs in the air, from the plant foliage, and from the ground surface.

The circular pattern of application makes it difficult to achieve a high coefficient of uniformity in delivering water to all plants. Sprinkler nozzles must be placed with sufficient overlap (between 40 to 60 percent) to achieve nearly uniform coverage of the surface area. More overlap is needed for windy situations.

Sprinklers usually throw water beyond the structure. Over-irrigation often occurs (i.e. the duration of the irrigation event is too long) since the containers on the fringes have to be adequately wet.

Many overwintering structures have a single row of sprinklers down the center. These sprinklers are much closer together in order to wet the interior area of the structure more uniformly. Containers are spread out under the part of the broadcast covered area where the water coverage is most uniform.

Sprinkler nozzles are designed to function at some pressure and discharge rate which determines the diameter of water throw. Operating pressure may range from 20 psi to 80 psi for large sprinklers. Discharge rates may range from 0.5 gallons per minute (gpm) to 1,000 gpm for a large gun. Wetting diameters range from 35 feet to 300 feet.

The application rate in inches per hour (iph) onto the ground is important. A rate of 0.2 iph to 0.5 iph is common. The crop should be watered fairly quickly to reduce the evaporative losses to the air and to reduce the time the foliage is wetted. At the same time, the irrigation should not cause runoff or puddles in fields. In container production an impervious surface may mean there will be runoff but it should be controlled.

It is a challenge to try to match sprinklers to any small or narrow area. In an overwintering (nursery) house, the sprinkler heads are placed close together in a line to get multiple overlaps of heads over some width of the structure. The goal is to achieve uniform coverage over the width of the structure. The many overlaps increase the application rate of the water and substrate wetting may suffer. Water applied too quickly may drain through quickly. Water will be thrown beyond the structure but the application rate will be low at the outer edge of the sprinkler pattern.

Overhead sprinklers are efficient for bedding plants, field crops, and small container plants placed close together. However, foliage is wet during each irrigation, so disease pressure may be a problem. Irrigations should be shceduled in time for the foliage to dry before relative humididty rises at night (in humid environments). In larger fields, the sprinklers can be placed in many rows so that overlapping of nozzles is possible and application uniformity can be good.

Overhead sprinklers include fixed installations, center pivots, traveling guns, and portable sprinklers. Each system has its own characteristics to consider.

Agricultural Water Management

Agricultural water management includes the management of water used in crop production (both rainfed and irrigated), livestock production and inland fisheries. Improved agricultural water management in these production areas is the answer to both global food security and poverty reduction.

The main challenge confronting water management in agriculture is to improve water use efficiency and its sustainability.

This can be achieved through (i) an increase in crop water productivity (an increased in marketable crop yield per unit of water transpired) through irrigation, (ii) a decrease in water losses through soil evaporation that could otherwise be used by plants for their growth, and (iii) an increase in soil water storage within the plant rooting zone through better soil and water management practices at farm and area-wide (catchment) scales.

Tracking and quantifying water fluxes at different spatial and temporal scales within the plant rooting zone remains a formidable challenge because of the interactions between water sources from rainfall, irrigation and subsurface water on plant uptake, soil evaporation, plant transpiration (water transpired by plants) and runoff or drainage losses from crop-growing areas. The use of isotopic and nuclear techniques to investigate the relative importance of soil and irrigation management factors that influence these interactions will greatly assist in the development of water management packages that involve the consideration of soil nutrient status, type of crops grown, growth stages and the overall agro-ecosystems to minimize not only water but also nutrient losses from the farmlands and enhance water and nutrient use efficiencies in agro-ecosystems under both rainfed and irrigated conditions.

Many nuclear and isotopic techniques are being employed in soil water management studies. The soil moisture neutron probe is ideal in field-scale rooting zone measurement of soil water, providing accurate data on the availability of water for determining crop water use and water use efficiency and for establishing optimal irrigation scheduling under different cropping systems especially under saline conditions.

The use of oxygen-18, hydrogen-2 (deuterium) and other isotopes is an integral part of agricultural water management, allowing the identification of water (and plant nutrients) sources and the tracking of water movement and pathways within agricultural landscapes as influenced by different irrigation technologies, cropping systems and farming practices. It also helps in the understanding of plant water use, quantifying crop transpiration and soil evaporation and allows us to devise strategies to improve crop production, reduce unproductive water losses and prevent land and water degradation.

References

- Safe-uses-of-agricultural-water: extension.psu.edu, Retrieved 16 May 2018

- Water-food-agriculture: lenntech.com, Retrieved 25 June 2018

- Irrigation-scheduling, potato: cropwatch.unl.edu, Retrieved 27 April 2018

- Irrigation-ditches-fact-sheet-1-201404141426: static.bouldercolorado.gov, Retrieved 09 March 2018

- Topic-water-management: naweb.iaea.org, Retrieved 23 May 2018

Chapter 4

Hydrozoning

The practice of growing plants with similar water requirements as a strategy of conserving water is known as hydrozoning. An elaborate study of the varied principles of hydrozoning has been provided in this chapter, which includes hydrozoning areas and benefits of hydrozones, among others.

A hydrozone defines an area of grouped plants with similar water needs. A hydrozone is assigned a water-usage level and a general irrigation type. This is useful for landscape architects who want to add a generalized irrigation element to a planting plan, for designers creating a water-efficient planting plan that conforms to regulations, or for irrigation professionals who are in the early stages of a design. By defining the water needs of a landscape design, you can estimate the total water usage of a site. This can be useful for regulation and certification purposes. Predefined worksheets can report hydrozone area, water use, and plants categorized by water needs.

Hydrozones can be broken into four categories.

(1) Very low hydrozones include plantings that need water when first planted, but none once established. Typical plants in this hydrozone include yarrow, rabbit brush and many native plants.

(2) Low hydrozones include plantings that generally do not require more than 3 gallons per square foot of supplemental water per year. During plant estab-

lishment or drought, additional supplemental water may be beneficial. Typical plants in this hydrozone include buffalograss, penstemon and daylily.

(3) Moderate hydrozones include plantings that generally require 10 gallons per square foot of water supplemental water per year. Typical plants in this hydrozone include turf-type tall fescue, potentilla and purple coneflower.

(4) High hydrozones include plantings that generally require 18 gallons of water per square foot of supplemental water per year. Typical plants in this hydrozone include Kentucky bluegrass, cottonwood, arborvitae and columbine.

Hydrozone Areas

Hydrozone areas can be broken down four zones:

Zone 1: Routine Irrigation. The principal hydrozone is the area that experiences both the greatest impact on the land and the largest water and energy use. For example, your backyard as a whole would be considered the principal hydrozone, because this is probably where you spend most of your time when you're outside.

Zone 2: Reduced Irrigation. Areas that are visually important but less used for activity are considered the secondary hydrozone. A good example is the shrub or flower bed near the main entrance of your home.

Zone 3: Limited Irrigation. Minimal hydrozones are the areas of your yard that receive little or no human use, and therefore justify little irrigation. These include buffer zones, distant views, and directional delineators such as strips of grass between the sidewalk and street and embankments. For best results and easiest maintenance, these areas should be matched with native plants that survive with pretty much only rainfall.

Zone 4: No Irrigation. The elementary hydrozone describes the area of your yard that receives only natural rainfall and no supplementary water supply. Here the human use intensity is lowest. These areas include spots for utilities, mulched parkways, and naturally existing vegetation.

Factors in determining zones range from types of plants in that area to element exposure. For example, grass, trees and flowers planted in an area that gets direct sunlight will require more water than those planted in a shady area. Similarly, if materials are planted at the top or bottom of a slope, irrigation should account for runoff and accumulation.

Even with careful planning of where you'll be placing plants, trees and shrubs based on element needs, you need to begin with a good base. Over time, soil compacts and forms a nearly impenetrable surface. To help ready your yard for water, give aeration a try. Aerating your yard breaks up the hard surface so the water can soak in to give your plants more oxygen, nutrients and water.

For planter beds, you can use a hand tool to gently turn the surface of the soil. Be careful working in the root zones so you don't damage your plants. For lawns, use a manual coring aerator (available at home improvement stores) or rent a machine – either one is easy to use. For best results, make two to three passes over each area, with holes about 3-inches apart.

Hydrozoning takes some strong up front planning but can allow for a healthy, sustainable landscape over time.

Benefits of Hydrozones

Keep your Turf

You can maintain appealing stretches of turf but control water use by restricting it to defined areas. By placing turf and other moisture-loving plants in the same zone, you streamline your watering needs and maximize the water you use.

Reduce Water Consumption

When hydrozoning is well conceived and properly executed, it can reduce landscape water use by 20% to 70%. The more area you devote to plants with low- to medium-water needs, the greater the savings.

Use Water more Efficiently

Because it allows you to tailor water use to the precise needs of the plants in each zone, hydrozoning maximizes every drop of water you use in your landscape.

Save Money

Reducing landscape water consumption allows you to save money on your water bill. If you live in a region where water rates are scaled to reward lower volume users, you may save even more just because you've reduced overall consumption.

Save Time

If you water all or some of your yard by hand, defined hydrozones will simplify your watering demands and reduce the amount of time you devote to the task.

Reduce the Need for Chemicals

Both over- and under-watering stress plants, turf and trees, promote disease and increase the need for fertilizers, chemical treatments and soil amendments. When plants get precisely the right amount of moisture, they tend to require fewer chemicals because they're stronger, healthier and more disease resistant.

Chapter 5

Environmental Impacts of Irrigation

Irrigation can have significant environmental impacts. These are manifested as a change in the quality and quantity of soil and water. It also results in alteration of the social and natural conditions downstream of an irrigation scheme. This chapter discusses the environmental impacts of irrigation, which includes direct and indirect effects, reduced river flow and increased groundwater recharge.

Direct Effects

Since irrigation systems deal with redirecting water from rivers, lakes, and underground sources, they have a direct impact on the surrounding environment. Some of these impacts include: increased groundwater level in irrigated areas, decreased water flow downstream of sourced rivers and streams, and increased evaporation in irrigated areas. Increased evaporation in irrigated areas can cause instability in the atmosphere, as well as increase levels of rainfall downwind of the irrigation. These changes to the climate are a direct result of changes to natural moisture levels in the surrounding atmosphere.

Indirect Effects

Irrigation systems also have an indirect impact on the surrounding environment. These indirect effects may not be as immediately noticeable as the direct issues. Additionally, these effects take a longer time to develop and produce longer-lasting changes. The indirect environmental impacts of irrigation include :

Waterlogging

When the conditions are so created that the crop root-zone gets deprived of proper aeration due to the presence of excessive moisture or water content, the tract is said to be waterlogged. To create such conditions it is not always necessary that under ground-water table should enter the crop root-zone. Sometimes even if water table is below the root-zone depth the capillary water zone may extend in the root-zone depth and makes the air circulation impossible by filling the pores in the soil.

The waterlogging may be defined as rendering the soil unproductive and infertile due to excessive moisture and creation of anaerobic conditions. The phenomenon of water-logging can be best understood with the help of a hydrologic equation, which states that

Inflow = Outflow -I- Storage

Here inflow represents that amount of water which enters the subsoil in various processes. It includes seepage from the canals, infiltration of rainwater, percolation from irrigated fields and subsoil flow. Thus although it is loss or us, it represents the amount of water flowing into the soil.

The term outflow represents mainly evaporation from soil, transpiration from plants and underground drainage of the tract. The term storage represents the change in the groundwater reservoir.

Causes of Waterlogging

After studying the phenomenon of waterlogging in the light of hydrologic equation main factors which help in raising the water-table may be recognised correctly.

They are:

I.   Inadequate drainage of over-land run-off increases the rate of percolation and in turn helps in raising the water table.

II.   The water from rivers may infiltrate into the soil.

III. Seepage of water from earthen canals also adds significant quantity of water to the underground reservoir continuously.

IV.   Sometimes subsoil does not permit free flow of subsoil water which may accentuate the process of raising the water table.

V.   Irrigation water is used to flood the fields. If it is used in excess it may help appreciably in raising the water table. Good drainage facility is very essential.

Effects of Waterlogging

The waterlogging affects the land in various ways. The various after effects are the following:

1. Creation of Anaerobic Condition in the Crop Root-Zone:

When the aeration of the soil is satisfactory bacteriological activities produce the required nitrates from the nitrogenous compounds present in the soil. It helps the crop growth. Excessive moisture content creates anaerobic condition in the soil. The plant roots do not get the required nourishing food or nutrients. As a result crop growth is badly affected.

2. Growth of Water Loving Wild Plants:

When the soil is waterlogged water loving wild plant life grows abundantly. The growth of wild plants totally prevent the growth of useful crops.

3. Impossibility of Tillage Operations:

Waterlogged fields cannot be tilled properly. The reason is that the soil contains excessive moisture content and it does not give proper tilth.

4. Accumulation of Harmful Salts:

The upward water movement brings the toxic salts in the crop root-zone. Excess accumulation of these salts may turn the soil alkaline. It may hamper the crop growth.

5. Lowering of Soil Temperature:

The presence of excessive moisture content lowers the temperature of the soil. In low temperature the bacteriological activities are retarded which affects the crop growth badly.

6. Reduction in Time of Maturity:

Untimely maturity of the crops is the characteristic of waterlogged lands. Due to this shortening of crop period the crop yield is reduced considerably.

Detection of Waterlogging

From the subject matter discussed above it is clear that the waterlogging is indicated when the ground water reservoir goes on building up continuously. When the storage starts building up in the initial stages the crop growth is actually increased because more water is made available for the crop growth. But after some time the waters table rises very high and the land gets waterlogged. Finally the land is rendered unproductive and infertile.

The problem of waterlogging develops in its full form slowly. Therefore its early detection is possible by keeping a close watch over the yields and also on the variations in the groundwater level. A comparative reduction in crop yields in spite of irrigation and fertilization and early maturity of crops indicate the symptoms of waterlogging. Also when harmful salts start appearing on the fields as white incrustation or deposit it indicates that waterlogging is likely to follow. In worst cases the water-table rises so high and close to the ground surface that the fields turn into swamps and marshes.

The best way of keeping watch over the problem of waterlogging is by observing variations in the groundwater level. It can be done by measuring the depth of water levels at regular interval in the wells dug in the area. Continuous high water levels indicate that the groundwater storage is building up which may create waterlogging in the area.

Solution to the Problem of Waterlogging

The problem of waterlogging may be attacked on two fronts. First is preventive measures, which keep the land free from waterlogging. Secondly curative measures may be adopted to reclaim the waterlogged area. But in principle both measures aim at reducing the inflow and augmenting the outflow from the underground reservoir.

Preventive Measures

(a) Controlling the loss of water due to seepage from the canals:

The seepage loss may be reduced by adopting various measures for example

I.   By lowering the FSL of the canal:

     Loss may be due to percolation or absorption but when FSL is lowered the loss is reduced to sufficient extent. It is course essential to see that while lowering the FSL command is not sacrificed.

II. By lining the canal section:

     When the canal section is made fairly watertight by providing lining the seepage loss is reduced to quite a good extent.

III. By introducing intercepting drains:

     They are generally constructed parallel to the canal. They give exceptionally good results for the reach where the canal runs in high embankments.

(b) Preventing the loss of water due to percolation from field channels and fields:

The percolation loss can be removed by using water more economically. It may also be affected by keeping intensity of irrigation low. Then only small portion of the irrigable tract is flooded and consequently the percolation loss takes place only on the limited area. It keeps the water-table sufficiently low.

(c) Augmentation of outflow and prevention of inflow:

It may be accomplished by introducing artificial open and underground drainage grid. It may also be achieved by improving the flow conditions of existing natural drainages.

(d) Quick disposal of rainwater:

Quick removal of rainwater by surface or open drains is a very effective method of preventing the rise in water table and consequent waterlogging of the tract. It is needless to state that the rainwater removed is net reduction in inflow.

Curative Measures

(a) Installation of lift irrigation systems:

When a lift irrigation project in the form of a tube well irrigation system is introduced in the waterlogged area the water table gets lowered sufficiently. It is found to be very successful method of reclaiming waterlogged land. Thus a combination of a canal system and a supplementary tube well irrigation system may be considered to be most successful and efficient irrigation scheme.

Of course it is true that it will create some complications while assessing the charges for irrigation water. (The canal water being cheaper than tube well water). Implementation of drainage schemes: The waterlogged area may be reclaimed by introducing overland and underground drainage schemes.

(b) Implementation of Drainage Schemes:

The waterlogged area may be reclaimed by introducing overland and underground drainage schemes.

Soil Salination

Salinity is the accumulation of salts (often dominated by sodium chloride) in soil and water to levels that impact on human and natural assets (e.g. plants, animals, aquatic ecosystems, water supplies, agriculture and infrastructure). Irrigation salinity occurs in irrigated landscapes.

Primary and Secondary Salinity

Primary (or inherent) salinity is the natural occurrence of salts in the landscape for example salt marshes, salt lakes, tidal swamps or natural salt scalds. Secondary salinity is salinisation of soil, surface water or groundwater due to human activity such as urbanisation and agriculture (irrigated and dryland).

Salt Sources

Salt may come from several sources including:

- Aeolian or wind borne salt from ocean spray or sedimentary deposits including dune sand and clay particles from the rivers and lakes of the Murray-Darling Basin.

- Cyclic salt from ocean spray or pollution dissolved in rainwater then deposited inland.

- Connate or fossil salt incorporated in marine sediments at the time of deposition, during periods when Australia was partly covered by sea.

- Rock weathering that allows salt to be released as minerals break down over time.

Hydrological Cycle

The hydrological cycle is the movement of water from the atmosphere to the earth and back again. Salts are highly soluble, so water is the key to the movement of salts in the landscape.

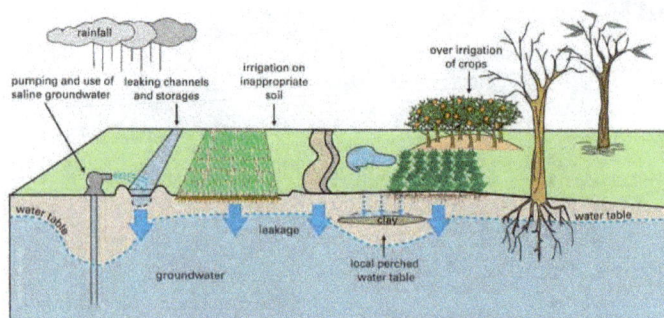

Groundwater System

The watertable is the surface below which all the spaces in soil and rock are filled with water. Water in this saturated zone is called groundwater. Above this is the unsaturated zone where the spaces are dry or only partially filled with water.

Water moving downwards past the plant root zone is called leakage. Water may leak from rivers, streams, dams and irrigation channels. Leakage that reaches the saturated zone is called groundwater recharge and groundwater that leaves the saturated zone is called groundwater discharge. Recharge areas are usually up-slope of discharge areas. When groundwater is at or near the soil surface discharge occurs as seepage, springs, and base flow to streams allowing groundwater to evaporate and/or be evapotranspired.

Causes of Irrigation Salinity

Irrigation salinity occurs due to increased rates of leakage and groundwater recharge causing the watertable to rise. Rising watertables can bring salts into the plant root zone which affects both plant growth and soil structure. The salt remains behind in the soil when water is taken up by plants or lost to evaporation.

Recharge rates in irrigation areas can be much higher than dryland areas due to leakage from both rainfall and irrigation. This causes potentially very high salinisation rates. Watertables within two metres of the soil surface indicate the potential for salts to accumulate at the soil surface.

Inefficient irrigation and drainage systems are a major cause of excess leakage and increase the risk of salinity and waterlogging in irrigation areas. Poor water distribution on paddocks results in some areas being under-irrigated, causing salts to accumulate (where watertables are high) and other areas being over-irrigated and waterlogged.

Groundwater mounds can develop under irrigation areas as a result of leakage from inefficient systems and restrictive layers. This puts pressure on the regional groundwater system forcing saline groundwater into waterways. Irrigating with saline water adds salt to the soil and increases the need for applying more irrigation water to leach salts past the plant root zone.

Salt Affects Plants through

1. Direct ion toxicity to plant tissue e.g. Leaf burn

2. Altered nutrient interactions e.g. Reduced availability of some elements

3. Influence on osmosis i.e. plants have difficulty extracting soil water.

Continual under-irrigation also increases salinity as salts contained in the irrigation water need to be flushed or leached periodically to prevent them accumulating to levels that limit productivity.

Inappropriate matching of crop, soil type and irrigation method can also cause excessive leakage. For example, irrigating high water-use crops using inappropriate irrigation methods should not be carried out on permeable soils (high sand content).

Other factors which influence leakage rates include soil type, climate and the amount (or removal) of deep-rooted perennial vegetation.

Replacing deep-rooted perennial pasture with irrigated annual crops reduces the level of evapotranspiration as rates are low following cultivation and during fallow periods. As a result, more water will infiltrate the soil profile and enter the water table.

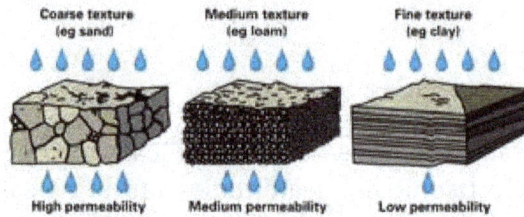

The permeability of different soil textures influences leakage rate to the groundwater system.

Effect of Salt on Plants

As salts accumulate in saline discharge areas they can reach levels that affect plants in a number of ways. This leads to poor plant health, a loss of productive species and dominance of salt-tolerant species.

Osmotic Effect

Under normal conditions, plants readily obtain water from the soil by osmosis (movement of water from a lower salt concentration outside the plant to a higher salt concentration in the plant). As soil salinity increases this balance shifts making it more difficult for plants to extract water.

Toxic Effect

Plant growth can be directly affected by high levels of toxic ions such as sodium and chloride. Excess sodium accumulation in leaves can cause leaf burn, necrotic (dead) patches and even defoliation. Plants affected by chloride toxicity exhibit similar foliar symptoms, such as leaf bronzing and necrotic spots in some species. Defoliation can occur in some woody species.

An excess of some salts can cause an imbalance in the ideal ratio of salts in solution and reduce the ability of plants to take up nutrients. For example, relatively high levels of calcium can inhibit the uptake of iron ('lime induced chlorosis'), and high sodium can exclude potassium.

Salinity Tolerance

Salt-tolerant plants (halophytes) can tolerate high internal concentrations of salts and take up salt with water. Examples include saltbush (Atriplex species) and bluebush (Maireana species).

Salt-resistant plants (glycophytes) cannot tolerate salt internally and exist in saline environments by excluding salt at their roots.

Most agricultural plants fall into the salt-resistant category of glycophytes. They can maintain growth in mildly saline soil by excluding salts at the roots. However, in extremely saline soils glycophytes are unable to both exclude salt and obtain sufficient water for maintenance. The impact of salinity varies with plant species, stage of growth, management practices, varieties and soil fertility.

Waterlogging exacerbates the effect of salinity. Waterlogged plant roots are unable to exclude sodium and chloride due to the increased rates of transport of these ions, and concentrations in the plant shoot increase. Poor aeration also affects soil biology responsible for converting nutrients to their plant available form, causing nutrient deficiencies.

Highly saline soils often become highly sodic. The ion imbalance and effect on the soil will depend on the type of salt present. Sodium and magnesium ions can destroy soil structure whereas calcium carbonate may improve soil structure (due to calcium) and increase soil pH (due to carbonate). Highly saline soils may have dark greasy patches where organic matter has been destabilised. On very salty sites a complete loss of groundcover and visible salt crystals often occur on the soil surface making it vulnerable to erosion.

Impacts of Irrigation Salinity

In the irrigation areas of NSW an estimated 4800 ha are affected by salinity .The areas most at risk include:

- Murrumbidgee Irrigation Areas near Griffith

- Murray Irrigation Areas around Deniliquin

- Jemalong Irrigation District west of Forbes.

The impacts of irrigation salinity are much the same as for dryland salinity. They include agricultural, environmental and social impacts.

Agricultural

Direct costs of increasing salinity to agricultural producers include:
- Reduced agricultural production
- Reduced farm income
- Reduced options for production
- Reduced access and trafficability on waterlogged land
- Reduced water quality for stock, domestic and irrigation use

- Damage to and reduced life of farm structures such as buildings, roads, fences and underground pipes and services
- Reduced productivity of agricultural land
- Animal health problems e.g. Saline water supply
- Farm machinery problems (bogging, rusting)
- Breakdown of soil structure, increased erosion and nutrient loss
- Loss of beneficial native flora and fauna
- Decreased land value.

Leaf burn in rice crops irrigated with highly saline water. Saline water application over 2 dS/m ECW will result in rice yield decline.

Environmental

Environmental Impacts from Land and Stream Salinity include:

- Decline of native vegetation and loss of habitat
- Loss of nesting sites and decline in bird populations
- Decline in wildlife fauna other than birds
- Reduced food for wildlife populations
- Increased soil and wind erosion
- Reduced wetland habitat and decline in fish and aquatic populations
- Reduced aesthetic value
- Reduced recreational and tourism values
- Reduced biodiversity in stream fauna, riparian vegetation and wetlands
- Increases in weeds and undesirable changes in plant populations
- Damage to state/national parks and wildlife sanctuaries

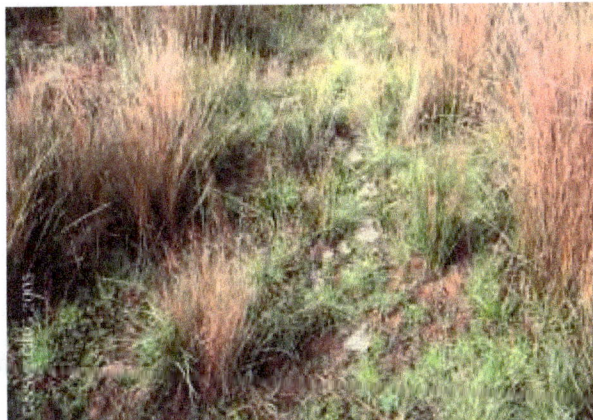

Species tolerant of waterlogging, for example Pin rush, become dominant.

Social

Impacts on the framework and structure of our society from increasing salinity include:

- Reduced aesthetic value of the landscape

- Reduced recreational and tourism values

- Reduced agricultural incomes due to productivity losses

- Flow-on impact on employment

- Reduced regional population (in both rural and urban communities)

- Increased pressure for consolidation of agricultural properties

- Reduced service levels to regional towns (especially <5000 persons)

- Increased social adjustment costs e.g. Welfare, marriage breakdown and bankruptcy.

Salinity induced by poor drainage and irrigation layout.

Increased Groundwater Recharge

Increased groundwater recharge stems from the unavoidable deep percolation losses occurring in the irrigation scheme. The lower the irrigation efficiency, the higher the losses. Although fairly high irrigation efficiencies of 70% or more (i.e. losses of 30% or less) can occur with sophisticated techniques like sprinkler irrigation and drip irrigation, or by well managed surface irrigation, in practice the losses are commonly in the order of 40% to 60%. This may cause the following issues:

- Rising water tables

- Increased storage of groundwater that may be used for irrigation, municipal, household and drinking water by pumping from wells

- waterlogging and drainage problems in villages, agricultural lands, and along roads - with mostly negative consequences. The increased level of the water table can lead to reduced agricultural production.

- Shallow water tables - a sign that the aquifer is unable to cope with the ground-water recharge stemming from the deep percolation losses

- Where water tables are shallow, the irrigation applications are reduced. As a result, the soil is no longer leached and soil salinity problems develop

- Stagnant water tables at the soil surface are known to increase the incidence of water-borne diseases like malaria, filariasis, yellow fever, dengue, and schisto-somiasis (bilharzia) in many areas. health costs, appraisals of health impacts and mitigation measures are rarely part of irrigation projects, if at all.

- To mitigate the adverse effects of shallow water tables and soil salinization, some form of watertable control, soil salinity control, drainage and drainage system is needed

- As drainage water moves through the soil profile it may dissolve nutrients (either fertilizer-based or naturally occurring) such as nitrates, leading to a buildup of those nutrients in the ground-water aquifer. High nitrate levels in drinking water can be harmful to humans, particularly infants under 6 months, where it is linked to "blue-baby syndrome"

Reduced River Flow

The consumptive nature of irrigation means that some change to the local hydrological regime will occur when new schemes are constructed and, to a lesser extent, when old schemes are rehabilitated. The ecology and uses of a river will have developed as a consequence of the existing regime and may not be able to adapt easily to major changes. It is also important to recognize the interrelationship between river flows and the water table. During high flow periods, recharge tends to occur through the riverbed whereas groundwater often contributes to low flows.

Low Flow Regime

Changes to the low flow regime may have significant negative impacts on downstream users, whether they abstract water (irrigation schemes, drinking supplies) or use the river for transportation or hydropower. Minimum demands from both existing and potential future users need to be clearly identified and assessed in relation to current and future low flows. The quality of low flows is also important. Return flows are likely to have significant quantities of pollutants. Low flows need to be high enough to ensure sufficient dilution of pollutants discharged from irrigation schemes and other sources such as industry and urban areas. A reduction in the natural river flow together with a discharge of lower quality drainage water can have severe negative impacts on downstream users, including irrigation schemes.

Habitats both within and alongside rivers are particularly rich, often supporting a high diversity of species. Large changes to low flows (±20%) will alter micro-habitats of which wetlands are a special case. It is particularly important to identify any endangered species and determine the impact of any changes on their survival. Such species are often endangered because of their restrictive ecological requirements. An example is the Senegal river downstream of the Manantali Dam where the extent of wetlands has been considerably reduced, fisheries have declined and recession irrigation has all but disappeared.

The ecology of estuaries is sensitive to the salinity of the water which may be determined by the low flows. Saline intrusion into the estuary will also affect drinking water supplies and fish catches. It may also create breeding places for anopheline vectors of malaria that breed in brackish water.

The operation of dams offers excellent opportunities to mitigate the potential negative impacts of changes to low flows.

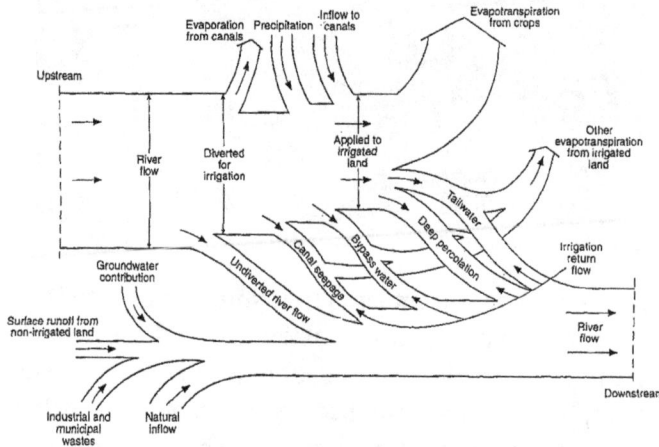

Irrigation return flow system for a given reach of a river system

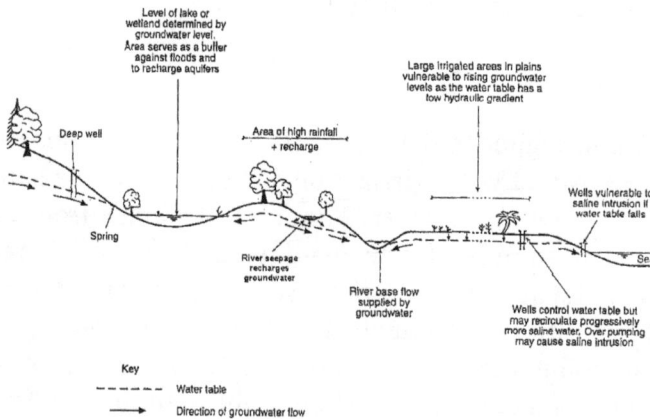

The interrelationship between surface water and groundwater

Flood Regime

Uncontrolled floods cause tremendous damage and flood control is therefore often an added social and environmental benefit of reservoirs built to supply irrigation water. However, flood protection works, although achieving their purpose locally, increase flooding downstream, which needs to be taken into account.

Radically altered flood regimes may also have negative impacts. Any disruption to flood recession agriculture needs to be studied as it is often highly productive but may have low visibility due to the migratory nature of the farmers practicing it. Flood waters are important for fisheries both in rivers and particularly in estuaries. Floods trigger spawning and migration and carry nutrients to coastal waters. Controlled floods may result in a reduction of groundwater recharge via flood plains and a loss of seasonal or permanent wetlands. Finally, changes to the river morphology may result because of changes to the sediment carrying capacity of the flood waters. This may be either a positive or negative impact.

As with low flows, the operation of dams offers excellent opportunities to mitigate the potential negative impacts of changes to flood flows. The designation of flood plains may also be a useful measure that allows groundwater recharge and reduces peak discharges downstream. This is one of the positive functions of many areas of wetland.

It is important that new irrigation infrastructure does not adversely effect the natural drainage pattern, thus causing localized flooding.

Operation of Dams

The manner in which dams are operated has a significant impact on the river downstream. There is a range of measures that can be undertaken to reduce adverse environmental impacts caused by changing the hydrological regime that need not necessarily reduce the efficacy of the dam in terms of its main functions, namely irrigation, flood protection and hydropower. Multi-purpose reservoirs offer enormous scope for minimizing adverse impacts. In the case of modifying low flows, identifying downstream demands to determine minimum compensatory flows, both for the natural and human environment, is the key requirement and such demands need to be allowed for at the design stage. The ability to mimic natural flooding may require modifications to traditional dam offtake facilities. In particular, passing flood flows early in the season to enable timely recession agriculture may have the added advantage of passing flows carrying high sediment loads.

A number of disease hazards are associated with dams some of which can be minimized, others eliminated by careful operation. They include malaria, schistosomiasis and river blindness.

Rooted aquatic weeds along the shore (or in shallow reservoirs) can be partially controlled by alternate desiccation and drowning. In some parts of the world local communities are willing to de-weed reservoirs and use the weeds as animal fodder.

Fall of Water Table

A possible advantage of reducing the water table level prior to the rainy season is that it may increase the potential for groundwater recharge. Lowering the water table by the provision of drainage to irrigation schemes with high water tables brings benefits to agriculture.

Lowering the groundwater table by only a few metres adversely affects existing users of groundwater whether it is required for drinking water for humans and animals or to sustain plant life (particularly wetlands), especially at dry times of the year. Springs are fed by groundwater and will finally dry up if the level falls. Similarly low flows in rivers will be reduced. Any changing availability of groundwater for drinking water supply needs to be assessed in terms of the economics of viable alternatives. Poor people may be disproportionately disadvantaged. They may also be forced to use sources of water that carry health risks, particularly guinea worm infection and schistosomiasis. In parts of Asia there are indications that lowering the ground level may favor the sandfly which may be vectors for diseases such as visceral leishmaniasis.

Saline intrusion along the coast is a problem associated with a falling groundwater level with severe environmental and economic consequences.

A continued reduction in the water table level (groundwater mining), apart from deleting an important resource, may lead to significant land subsidence with consequent damage to structures and difficulties in operating hydraulic structures for flood defence, drainage and irrigation. Todd gives an example of a drop in ground level of over 3 m associated with a 60 m drop in groundwater level over a period of 50 years in the Central Valley, California. Vulnerable areas are those with compressible strata, such as clays and some fine-grained sediment. Any structural change in the soil is often irreversible. The ground level can fall with a lowering of the water table if the soils are organic. Peats shrink and compact significantly on draining, with consequent lowering of the ground level by several meters.

Particular care is needed in the drainage of tropical coastal swamp regions as the $FeSO_4$ soils can become severely acidic resulting in the formation of "cat-clays".

A number of negative consequences of a falling water table are irreversible and difficult to compensate for, eg salt water intrusion and land subsidence, and therefore groundwater abstraction needs controlling either by licensing, other legal interventions or economic disincentives. Over-exploitation of groundwater, or groundwater mining, will have severe consequences, both environmental and economic, and should be given particular importance in any EIA.

Rise of Water Table

In the long-term, one of the most frequent problems of irrigation schemes is the rise in the local water-table (waterlogging). Low irrigation efficiencies (as low as 20 to 30% in some areas) are one of the main causes of rise of water table. Poor water distribution systems, poor main system management and archaic in-field irrigation practices are the main reason. The ICID recommendation to increase field application efficiency to even 50% could significantly reduce the rise in the groundwater. The groundwater level rise can be spectacularly fast in flat areas where the water table has a low hydraulic gradient. The critical water table depth is between 1.5 and 2 m depending on soil characteristics, the potential evapotranspiration rate and the root depth of the vegetation/crops. Groundwater rising under capillary action will evaporate, leaving salts in the soil. The problem is of particular concern in arid and semi-arid areas with major salinity problems. A high water table also makes the soil difficult to work.

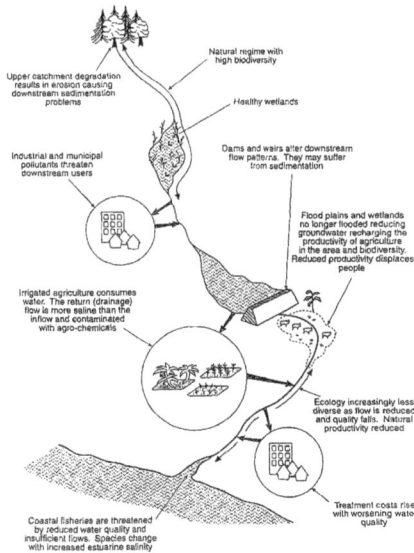

Causes and impacts of reduced water quality in a river system

Good irrigation management, closely matching irrigation demands and supply, can reduce seepage and increase irrigation efficiency, thereby reducing the groundwater recharge. The provision of drainage will alleviate the problem locally but may create problems if the disposal water is of a poor quality. Apart from measures to improve water management, two options to reduce seepage are to line canals in highly permeable areas and to design the irrigation infrastructure to reduce wastage. Waterlogging also implies increased health risks in many parts of the world.

References

- N.T. Singh, 2005. Irrigation and soil salinity in the Indian subcontinent: past and present. Lehigh University Press. ISBN 0-934223-78-5, ISBN 978-0-934223-78-2, 404 p

- What-is-the-environmental-impact-of-irrigation: worldatlas.com, Retrieved 11 May 2018

- Rosenburg, David; Patrick McCully; Catherine Pringle (2000). "Global-Scale Environmental Effects of Hydrological Alterations: Introduction". BioScience. Sep 2000: 746–751. doi:10.1641/0006-3568(2000)050[0746:GSEEOH]2.0.CO;2. Retrieved 17 March 2014

- Waterlogging-definition-causes-effects-with-statistics-61000: yourarticlelibrary.com, Retrieved 31 March 2018

- "Irrigation potential in Africa: A basin approach". Natural Resources Management and Environment Department. Retrieved 13 March 2014

- Irrigation-salinity-causes-and-impacts-310365: dpi.nsw.gov.au, Retrieved 12 July 2018

- O. A. Tuinenburg et al., The fate of evaporated water from the Ganges basin, Journal of Geophysical Research: Atmospheres, Volume 117, Issue D1, 16 January 2012

- Bruce Sundquist, 2007. Chapter 1- Irrigation overview. In: The earth's carrying capacity, Some related reviews and analysis. On line : "Archived copy". Archived from the original on 2012-02-17. Retrieved 2012-02-17

Chapter 6

Irrigation Technology

The study of irrigation technology is vital for a holistic understanding of irrigation. This chapter closely examines the different aspects and uses of valley variable rate irrigation, site-specific management, GPS guidance, T-L irrigation, rain sensor, etc. in agriculture.

Changes and advances in the irrigation industry have led to new technologies.

Agriculture technology is enabling farmers to be more in control of what's happening in their fields and with their equipment. Variable rate seeding allows precise efforts based on soil variability. Newer sprayers feature individual nozzle control to get the precise drop size and amount per crop or location in a field. The data provided by UAV technology is giving farmers the opportunity to spot potential problems while they're still minor enough to fix.

Examples of irrigation technology include:

Reinke

Reinke's new computerized irrigation management system gives growers more control over essential functions from their office or laptop computers. The Reinke Automated Management System (RAMS) provides increased crop production control with custom water and chemical application program capabilities and helps conserve energy, water and labor costs.

"RAMS is the best application choice to meet precise farming demands in the 21st century," said Tim Goldhammer, Reinke vice president of marketing.

A new Temperature Control Transmitter also protects systems from a hard freeze. The transmitter is accurate to one degree and will automatically shut off an irrigation system in freezing temperatures.

When incorporated with RAMS, the Temperature Control will give the operator the ability to set the desired shutdown and restart temperatures through the program software.

"All control settings are executed through keystrokes on the panel and are easily changed and/or monitored," said Tim Zikmund, Reinke technical products manager.

The Temperature Control also works with a non-computerized standard panel configuration. Temperatures are set through a manually operated dial inside the main control panel.

RAMS features a chemical pump status report and also measures system control increments to a tenth of a degree for efficient water management. The Reinke Phone Link allows a grower to check or change status of the system from a landline telephone or cell phone at any time. Phone Link will also contact the grower if a problem arises with the system.

Reinke's system diagnostics result in less troubleshooting and down-time," Zikmund said.

Telemetry control incorporates UHF radio technology to monitor, operate and compile detailed irrigation records of farm operations for up to 100 systems.

Valley Variable Rate Irrigation

With a combination of hardware and software, patented Valley Variable Rate Irrigation (VRI) allows you to customize water application based on topography information, soil data maps, yield data and other user-defined information. Based on your VRI Prescription, you're applying water only where it needs to be. So you're not applying water to unnecessary areas in your field, such as ditches, canals, buildings and boggy areas.

VRI is precision agriculture made easy. You can wirelessly upload your VRI Prescription to your center pivot with Valley and AgSense remote communication products. And because you can wirelessly upload your prescription, you can quickly make changes as your water application needs change in the field. Don't use remote communication? You can also hard-wire upload your prescriptions directly to your computerized control panel.

Advantages of VRI

- Make all areas in your field more profitable.

- Reduce runoff, which can help your soil's health and the environment.

- Increase water and chemical application efficiency.

- RELAX — you know that your precious water resources are being used wisely.

- Valley precision irrigation equipment enhances your agricultural productivity.

T-L Irrigation

T-L's newly patented technology has successfully married center pivot technology with drip irrigation. PMDI consists of in-line drip hoses spaced at 60 inches between lines dragged through various crops by a center pivot or linear move irrigation system. PMDI combines the efficiency of surface drip irrigation (95 percent) with the flexibility and economics of center pivot irrigation.

PMDI consists of in-line drip hoses spaced at 30 inches, 40 inches, and 60 inches between lines being dragged through various crops by a center pivot or linear move irrigation system.

PMDI combines the efficiency of surface drip irrigation (95 percent) with the flexibility and economics of center pivot irrigation. There are two main advantages of Precision Mobile Drip Irrigation.

- Over-all water efficiency. Since the drip line is on the ground, you eliminate evaporation and wind drift associated with traditional sprinklers because wind will not affect it as it applies the water at a super efficient 95 percent. You get all the efficiency of surface drip at the much-reduced cost per acre price of center pivots.

- Dry wheel tracks. In many soils and cropping practices, deep wheel tracks on pivots and linears are a problem. With PMDI the drip lines water behind the wheels so the tires run on dry ground.

Rain Sensor

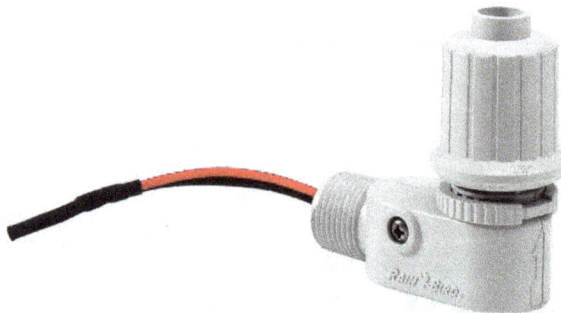

A rain sensor is an irrigation shutoff device that prevents an automatic sprinkler system from turning on during and after a rain storm. Rain shut-off sensors are wired to a sprinkler timer and override the scheduled irrigation when a sensor on the shutoff device detects water.

When the collected rainwater has evaporated from the sensor, scheduled irrigations resume. Rain shut-off sensors are simple, economical and useful tools for preventing irrigation that would be wasteful. Rain shut-off sensors work best for short off periods.

For extended periods, it is more accurate to have sprinkler timer in the "off" position. City of Santa Barbara Water Customers can get free rain shut-off sensors.

GPS Guidance

The development of control technology in precision mechanized irrigation has led to the application of GPS technology. GPS (Global Positioning System) was created and realized by the American Department of Defense (DOD), and was originally based on and operated with 24 satellites (21 required satellites and 3 replacement satellites). Today, approximately 30 active satellites orbit the earth from a distance of 20,200 km. From either the earth's atmosphere or low orbit, GPS satellites transmit signals which find the exact location of a GPS receiver; the receiver must be on the surface of the earth to acquire the GPS coordinates. GPS is currently being used in aviation, nautical navigation, and determining position on land. Furthermore, it is used in land surveying and other applications where the determination of an exact position is required. GPS is a free service that can be used by any person in possession of a GPS receiver; the only requirement is an unobstructed view of the satellites.

Agricultural applications use GPS technology for equipment guidance; the equipment uses lightbar-guided and automated steering systems that help maintain precise swath-to-swath widths. Guidance systems are packaged with a display module that issues audible tones or lights as directional indicators for the operator's use. The operator monitors the lightbar to maintain the desired distance from one swath to the next. Automated steering systems integrate GPS guidance capabilities with the vehicle steering system.

GPS is also used with yield monitoring systems; the sensor is typically located at the top of the clean grain elevator. As the grain is transported into the grain tank, it strikes the GPS sensor and the amount of force applied to the sensor represents the recorded yield. The data is displayed on a monitor located in the combine cab and stored on a computer card that is transferred to an office computer for analysis.

Another use for GPS in agricultural applications is field mapping in which it is used to locate and map specific field regions, such as areas that include high weed, disease, and pest infestations. Objects such as rocks and poorly drained regions can be recorded as landmarks for future reference. GPS is used to locate and map soil sampling locations, allowing growers to develop contour maps that show fertility variations throughout fields.

GPS has also been commonly used in agriculture for precision crop input applications. The technology is used to vary crop inputs throughout a field based on GIS maps or real-time sensing of crop conditions. Variable rate technology requires a GPS receiver, a computer controller, and a regulated drive mechanism mounted on the applicator. Crop input equipment, such as planters or chemical applicators, can be equipped to vary one or several products simultaneously. As GPS technology has been used in agriculture for several years, it is only logical that it should also be applied to mechanized irrigation.

With center pivots, control technology has been used for more than thirty years to stop the machine, reverse it automatically, and turn endguns on and off. These controls were originally based on electro mechanical switches stationed at either the last regular drive unit or at the pivot point. The function of these switches depended on their physical placement; it was often difficult to estimate how the pivot would behave, particularly when turning endguns on and off. Depending on certain circumstances, the endguns could turn on or off fifty feet or more from where the operator intended.

In the early 1990s, new position methods were developed by the center pivot manufacturers to provide position information to computerized control panels. These included, but were not limited to, resolvers and encoders. The position information was displayed on the panel, enabling the operator to determine settings that controlled endguns and stopping of the pivot. The computerized control panel then made the positioning decisions instead of trying to rely on mechanical switches. With the advent of computerized control panels, the door opened for multiple changes around the field, such as the control of six pie shaped sectors or more, multiple settings on a single endgun, and control of a second endgun. These innovations moved mechanized irrigation into the realm of precision irrigation. If setup properly, these control panels were reasonably close to duplicating what the operator wanted; but, like the aforementioned mechanical switches, they were still using a positioning estimate at the pivot point, a large drawback. At this time, a decent solution for linear machine positioning did not exist.

In the past, the most successful guidance solution for center pivot corner arms depended on following a signal from a buried wire. Successful guidance for linears had three choices – following a furrow near the cart, following an aboveground cable (also near the cart), or following a signal from a buried wire (similar to the center pivot with a corner arm), which was usually placed near the middle of the linear machine. It was found that each of these guidance solutions have limitations due to the high risk of damage from farm operations and/or lightning.

Mechanized irrigation manufacturers have been working on GPS applications since early 2000 the first commercial packages utilizing GPS for precision irrigation control arrived on the market a few years later. The early GPS applications were first focused on center pivots and secondly on linear machines in order to eliminate the need for cables and/or any other type of land-based guidance, such as furrow.

Work rapidly expanded to use GPS in providing position information to a center pivot as an alternative to electro mechanical devices reporting position at the pivot point. The GPS receiver is placed on the last regular drive unit and is controlled either by sending information to the computerized control panel or by sending the information directly to another control device. Market suppliers quickly entered the field, such Farmscan, which soon began to utilize GPS information to control banks of sprinklers based on a pre-determined prescription map.

Current applications of GPS with mechanized irrigation include reporting position for center pivots and linear machines, and guidance of linears and center pivots with corner arms.

Center pivots that utilize GPS technology on the last regular drive unit can replace previously used mechanical switches, resolvers, and encoders, which estimate position information from the pivot point. As the GPS receiver is stationed at the last regular drive unit, the technology provides more accurate information on the position rather than estimating the position of the last regular drive unit from the pivot point. Estimating from the pivot point has the potential for errors, unless the pivot alignment is maintained in an extremely straight position. The more spans on the center pivot, the more the risk for error due to nonalignment. Depending on the center pivot manufacturer, the grower may configure and adjust endguns or pivot stops using the GPS information found either at the computerized control panel or at the end of the center pivot with a PDA or laptop through Bluetooth technology. Third-party suppliers of GPSbased units operate independently from the control panel, providing information via the internet. Depending on the supplier, the GPS data can be used by the control device to program the on and off cycles for one or two endguns, stop, reverse, and change the speed of the center pivot, or, depending on the sprinkler hardware, turn on or off banks of sprinklers.

GPS technology is also being used for linear applications to control endguns and the machine's operation, which includes stopping, reversing, and changing speed. When utilized with linears, the GPS data is processed in a specially designed control panel, such as the Valley AutoPilot Linear.

On both the center pivot and linear GPS position, accuracies of +/- 3 meters is typically recorded with single-band GPS receivers and WAAS (Wide Area Augmentation System) correction signal. Guidance for linear machines and center pivots with a corner arm is different from positioning in that the desired accuracy must be much higher in order to maintain correct tracking of the linear or corner. Generally, the accuracy is typically +/- 3 cm. To achieve this accuracy, the addition of a reference base station is required.

The marketplace has rapidly embraced the use of GPS technology in agricultural equipment, such as tractors, combines, and sprayers. GPS has literally become a part of the farming life. The acceptance of GPS for mechanized irrigation has begun to develop and flourish, and farmers expect GPS technology to be utilized with irrigation. Growers who currently use GPS for positioning linear machines and center pivots are pleased with the performance and the ease of operation. For instance, one farmer recently stated that his operators used to complain when the air conditioning went out of the tractor cabs, but now the first thing they complain about is when the GPS is malfunctioning.

Still today, some equipment manufacturers offer control panels without GPS to provide a traditional technology choice to their customers. The cost of GPS for center pivot or linear positioning varies with the type of package that best suits the customer's needs. Costs can be three to six times the price of an encoder or resolver type position sensor; however, the improved accuracies shown and recorded outweigh the high price tag.

GPS guidance for linear machines has good reception and acceptance by growers, the only drawback being the perceived cost. Customers using linear GPS guidance have experienced much improved performance and accuracies; this is because a linear machine always tries to maintain itself perpendicular to whatever guidance is being used, and the machine will continually steer itself to accomplish this. Each of the traditional types of linear guidance – furrow, aboveground cable, and below-ground cable – is difficult to install and/or maintain perfectly straight. As more steering is needed to keep alignment straight using traditional guidance solutions, there is more potential for non-uniform watering and delays moving down the field. Customers who have switched to GPS guidance have observed significantly less steering and more consistently completed field passes, which inevitably lead to better watering patterns and more dependable operation of the linear. The cost of GPS guidance varies greatly when compared to below-ground cable guidance, as in some cases, due to installation costs, the GPS guidance may actually be less expensive. Another major benefit of GPS guidance is it cannot be damaged due to lightning; in some areas of the United States, below-ground guidance cannot be used at all due to constant occurrences of sky to ground lightning.

Site-specific Management

Going hand-in-hand with the opportunities of VRI systems and GPS, technology has greatly impacted the ways in which irrigation systems are managed.

Original control panels were developed in the early 90s to work in conjunction with VRI systems. They allow a farmer to set the rate of application and adjust at any time from anywhere, without having to go into the field.

Modern control panels now enable variable rates across the same field; for example, one half receiving ½ inch of water, while the other half receives a full inch. They can also be programmed to auto-adjust the system based on temperature, set to allow the pivot to travel a specified number of degrees in a designated time period, and programmed months in advance, all directly from the panel or even a smartphone.

Not just applicable to VRI, virtually every system on the farm, including pumps and sensors, can be monitored and managed through any device, from a Smartphone to a tablet to a desktop computer. Best of all, these systems record and store all the data they collect for future use and decision-making.

As for what can be managed, the possibilities are virtually endless. Programs are available to provide crop monitoring and measurement of soil moisture content. Others enable control of a pivot's starting, stopping, direction, and speed. Some can be paired with weather sensors or other programs to further enhance their capabilities and precision. And many of them track real-time info and send alerts to the device, letting a farmer know if something should be looked into.

Chapter 7

Irrigation Systems

There are different kinds of irrigation systems, such as manual, gravity, lift, sewage irrigation systems, etc. This chapter delves into the varied aspects of these systems to provide a holistic understanding of irrigation.

Manual Irrigation

Manual irrigation systems are easy to handle, require no technical equipment and are therefore generally cheap (in contrast to high-tech systems such as sprinkler irrigation or subsurface drip irrigation). But they need high labor inputs. A common and very simple technique for manual irrigation is for instance the use of watering cans as it can be found in peri-urban agriculture around large cities in some African countries. A more sophisticated and very water-efficient type of manual irrigation system is small-scale drip irrigation with buckets Beside these systems, there are many other methods for manual irrigation, which are easy to install and simple to use. In general, all of these methods have high self-help compatibility and a relatively high performance. Therefore such systems are also called helpful irrigation methods: high frequency, efficient, low-volume, partial-area, farm-unit, and low-cost.

Basic Manual Irrigation Principles

Low-Cost Drip Irrigation System

Simple drip irrigation (in contrary to high tech drip irrigation systems) uses low-cost plastic pipes cut to the appropriate lengths laid on the ground to irrigate vegetables, field crops and orchards. Small holes in the hose allow water to drip out and keep the base of the plant wet without wasting any water.

A low-cost "farm-kit system" with a 1000 litres water tank can service up one-eight of an acre.

Watering Cans

Irrigation by watering cans is a very basic way but is still widely used. This creates a lot of work for the labors especially if this technique is used for large fields. A common way to make this work easier is a carry-pole across the shoulders. The field worker is able to carry two big watering cans in each side and the irrigation water can de distributed equally on the field. A rose can be added to the watering-can to create a "sprinkler effect".

With watering cans, the field worker is able to irrigate very specific and only where it is necessary.

Pitcher Irrigation

A very basic subsurface method consists in placing porous clay jars (or pots) in shallow pits dug for this purpose. Soil is then packed around the necks of the jars so that their rims protrude a few centimetres above the ground surface. Water is poured into the jars either by hand or by means of a flexible hose connected to a water source. Since the walls of the pots are porous (make sure to use unglazed pots), the water can seep slowly out and reach the roots of the plants. The jars can be made of locally available clay: they are of no standard shape, size, wall thickness or porosity. Instead of a clay or earthenware pod, also the sweet monkey orange fruit (Strychnos spinosa) can be used when it has been dried and the top cut off. The jar should be filled up regularly (especially in arid areas) and has to be changed if there are big cracks that the water percolates without reaching the roots.

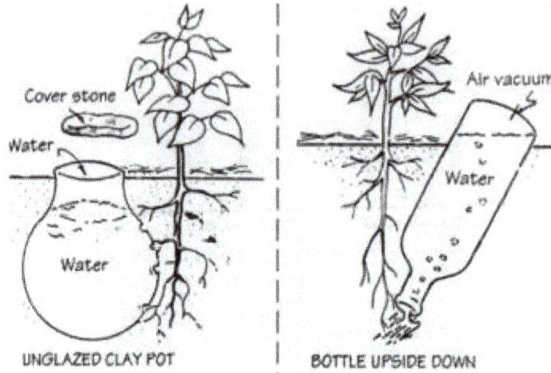

Left: Clay pot irrigation method. Right: Bottled irrigation method is also effective and simple. Bottles can be found everywhere in the world.

Bottle Irrigation

Similar to the pot method described above, pitcher irrigation can also be done using a bottle. The bottle is first filled and than placed with its neck into the soil next to a plant, so it stands upside down. The dense soil hinders the water from leaving the bottle immediately. Instead, it gets released slowly and directly besides the roots, so it is available to the plant for a longer time and the water cannot evaporate directly.

Porous and Sectioned Pipes

Another variation of pitcher irrigation uses porous pipes instead of pots to spread water along a continuous horizontal band in the soil, rather than at discrete locations. As such, the porous pipe method is more suitable for closely spaced row crops grown in beds, such as vegetable crops. One end of the porous pipe is made protrude above ground that the operator is able to refill it with water as soon as it is necessary. In contrast to subsurface drip irrigation, the porous pipe distributes the water over the whole length and not only where it is perforated. That means it is less effective and water loss is slightly higher.

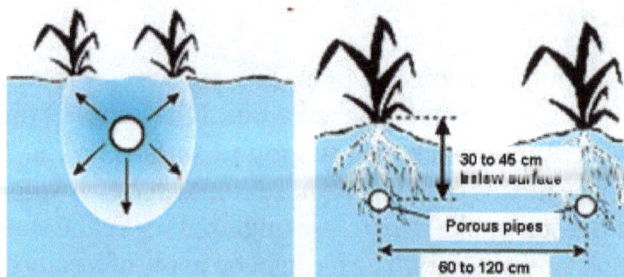

Schematic view of porous pipe irrigation.

Perforated Plastic Sleeves

An interesting variation of the pitcher method is the use of thin plastic sheeting to form a sleeve-like casing. To define its comparative usefulness better, the method should be tested side by side with alternative methods of irrigation. To date, this has not been done systematically.

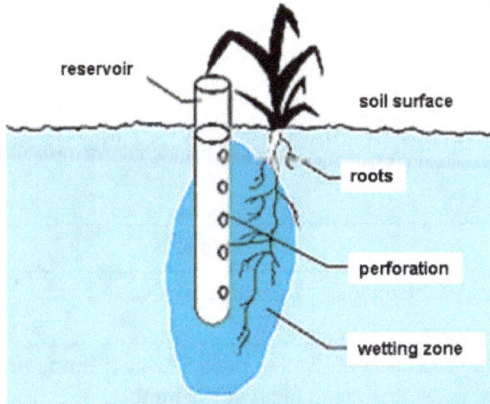

The plastic sleeve method is not tested systematically and therefore it is difficult to estimate its performance

Costs Considerations

All of the described systems are very cheap. Many of them can be made out easily available material (e.g. old buckets or bottles) or renewable resources (bamboo). This reduces the costs significantly. However, high labour inputs are required for operation and maintenance.

Operation and Maintenance

As water is brought into the system manually, this requires high labour input. Moreover, it is important to check the systems regularly to prevent blockages and leakages. If there are any problems it should be cleaned and/or fixed as fast as possible to prevent damages on crops. Furthermore, there are several techniques to improve the production and avoid water loss on the fields.

Health Aspects

If wastewater is used for the manual irrigation process, there are potential health risks if water is not properly pre-treated (i.e. inadequate pathogen reduction). If poorly treated wastewater is applied. Appropriate pre-treatment should precede any irrigation scheme to limit health risks to those who come in contact with the water. As well, depending on the degree of treatment that the effluent has undergone, it may be contaminated with the different chemicals that are discharged into the system. When effluent is used for irrigation, households and industries connected to the system should be made aware of the products that are and are not appropriate for discharging into the system. Drip irrigation and subsurface drip irrigation are the only types of irrigation that should be used with edible crops, and even then, care should be taken to prevent workers and harvested crops from coming in contact with the treated effluent. Despite safety concerns, irrigation with effluent is an effective way to recycle nutrients and water

Applicability

Manual irrigation methods are appropriate for small-scale farming or backyard gardening irrigation in dry and arid climates where water is scarce. All the different designs reduce water evaporation. It allows people to grow its own food with simple but effective techniques.

Advantages	Disadvantages
Improved water-use efficiency (reduced loss through evaporation)	Labour intensive
Well directed, selective and targeted irrigation	User need a basic training to install and use the correct most of the method
Ensures constant water supply in the crucial phase of germination	If the water is not properly filtered and the equipment not properly maintained, it can result in clogging
Higher yields, better quality, higher germination rate, lower incidence of pest attack	Manual subsurface drip irrigation avoids the high capillary potential of traditional surface applied irrigation, which can draw salt deposits up from deposits below
Facilitates pre-monsoon sowing	
Can be constructed with locally available material	
Low investment costs	

Gravity Irrigation

Gravity irrigation is a method of irrigation in which you can water plants anywhere, in

the ground or in containers, even if there's no electricity available. These systems work by taking advantage of water's ability to travel the path of least resistance all on its own. You don't need any pumps and can use this for watering your garden temporarily or all season long. It is best used on individual plants, small clusters, or row crops.

Without a pump, you need a reservoir. You can fill it with collected rainwater, from the hose, or transported from nearby a pond or stream. The water moves to your plants' roots through drip irrigation. Just don't forget to check your water level. It will run out faster than you expect during a drought or a dry climate, especially as your plants grow larger. You can also add fertilizer or nutrient supplements to your irrigation reservoir.

Since delivery of gravity irrigation is via drip rather than spray, it conserves water. And if constructed right, you can use it for sub-irrigation, reducing evaporation loss even more. However, below the surface drip irrigation might call for a reservoir for every plant. It depends on the size of your plants and your reservoir.

You can buy gravity irrigation kits and systems, but it's easily turned into a DIY irrigation project. You can use water or soda bottles for individual plants. To water a row or group of plants from one drip irrigation reservoir it's best done using buckets, totes, or barrels with a tight-fitting lid. You don't want to lose water to evaporation from the container. It's also important to stop debris from collecting that can plug up your drip system.

One ancient gravity irrigation system is known as an Olla pot. These clay vessels buried in the middle of a ring of plants drip water through the porous sides whenever the water pressure around it drops too low to hold it in. You can use one alone or create a connected system that automatically refills a chain of them from a reservoir. Garden shops sell the clay versions, but a plastic drink bottle works well too. Drill tiny holes down one or all sides and bury it next to plant roots. Using the cap end for refills works better than cutting off the bottom. Gardeners use them in flower pots, container gardens, and traditional plots. However, you need a larger reservoir for large plants and multiple plant like a big jug.

Setting up gravity irrigation that delivers water at the surface calls for elevating your reservoir. The simplest setup makes use of 1-2 liter PET bottles with either hole drilled in the cap or purchasing spike emitter caps. Since this leaves your reservoir exposed to sunlight, it's best to use transparent bottles as a temporary solution while you're out of town. Because if the sun can penetrate it algae will grow in your reservoir. For long-term use, paint repurposed drink bottles a dark color.

Got an army of containers? A raised bed or backyard garden? Consolidating your DIY irrigation to draw from a larger drip irrigation reservoir might make far more sense. Here you will want to add a hose timer to manage water flow because, with the water reservoir raised off the ground, the flow won't shut off by pressure in the soil. And you do want it elevated to improve your water pressure unless you only need it to travel short distances.

One urban farming nurse's solution in a cucumber patch was to incorporate IV drip lines used in hospitals. This would allow some special individual plant treatment and will work in a pinch, but perhaps isn't suited to all water sources. Hose or tap timers aren't always expensive and come in manual, solar, and battery-powered models. Just connect it to the spigot tapped into your raised reservoir and attach the main water line feed to the timer outlet.

In such a DIY irrigation system you can connect both weeping drip hose or drip line tubing delivering water down your rows or grid. Using drip line tubing allows you to choose where water is emitted. So, you use water only on your plants, which increases your water conservation. You can also attach spaghetti line and drip emitter caps to increase water delivery to larger root systems. However, drip emitters and spaghetti line can easily clog from debris and particles in pond, stream, and well water. If any of those are your water source, look for drip system parts rated for dirty water.

Gravity irrigation isn't just for outdoor gardening. In fact, a system like the AutoPot found at indoor gardening shops uses gravity fed sub-irrigation made available to the plants through capillary action. Here a specially designed mechanism controls water flow, triggered only when the growing media is dry.

You will also see people using the pop bottle irrigation reservoir for indoor plants. Some partially or fully bury them next to the plant. Others make use of the drilled or spike emitter cap. And then there's the primed line drip trick. Filling a jar or jug with water and placing it raised above the plants. Next, create your gravity irrigation via a piece of tubing much like siphoning gas. Suck the water to the end and stick it in the growing media. It will only let water into the container when depleted water pressure allows it to do so. Note that you need small tubing or you'll have more than a drip action.

There's an automatic irrigation option that fits anyone's style of gardening, no matter where on Earth you live. Gravity irrigation can keep plants happy and healthy with a fine-tuned manufactured system, a simple conversion kit, or full-blown DIY irrigation for a big garden or farm that need multiple reservoirs. It's not always necessary to have fancy stuff to grow well. All your veggies care about is consistent moisture, lots of sunshine, and a balanced diet of quality nutrients.

Lift Irrigation

Lift irrigation is a method of irrigation in which water instead of being transported by natural flow (as in gravity-fed canal systems) requires external energy through animal, fuel based or electric power using pumps or other mechanical means. Treadle pumps, although an ancient method of lifting water for small heads have recently been modernized and used in a big way.

Lift irrigation schemes must accomplish two main tasks: first, to carry water by means of pumps from the water source to the main delivery chamber, which is situated at the top most point in the command area. Second, they must distribute this water to the field of the beneficiary farmers by means of a suitable and proper distribution system. The source is mainly groundwater, river streams, contour canals, ponds and lakes.

For a viable lift irrigation scheme, the requirements are constant water source should for the whole irrigation season at the site and the feasibility to lift water to the desired location. Different capacity pumps are required depending upon the duty point head, and discharge. The rising main may be of steel, concrete or any other suitable material. Lift irrigation schemes are useful where the target land is at higher level.

The advantage of lift irrigation is the minimal land acquisition problem and low water losses. The lift irrigation schemes are instrumental in stabilizing agriculture production particularly in the years of droughts and increase food production as water is available whenever it is required and thereby increase in income level.

Lift irrigation schemes are either individually owned or owned by a group of farmers in a cooperative mode. For successful functioning the lift irrigation schemes require appropriate technique, planning, designing and execution through knowledgeable technical person. Participation of beneficiaries is quite necessary. Unplanned development of lift irrigation systems have the potential to have its adverse impact on the groundwater levels, as has been the case in many south Asian countries in the recent years. Continuous drop in groundwater table is making the cost of running and maintenance of lift irrigation schemes more costly. Cooperative lift irrigation schemes have the potential to be participatory in development and management. Multistage submerged pump is at the heart of any irrigation system.

Lift Irrigation on Surface Water

Size and Location

The cultural command area should be restricted up to 20- 40 ha only under small lift irrigation schemes. Lift irrigation schemes on surface water source with command area more than 40 ha may be considered as medium / large schemes and the technical and financial appraisal may be carried out as per the guidelines pertaining to large lift irrigation schemes.

The scheme area should be as close to source of water as possible and land holdings are contiguous as far as possible from design optimization point of view.

Command Area / Cropping Pattern

The command area has to be demarcated and as far as possible to be adjoining. The cropping pattern has to be fixed and water requirement has to be estimated.

Design Discharge

The design discharge for small lift irrigation scheme should be assessed based on peak water requirements for envisaged cropping pattern and area of crops during kharif, rabi and summer seasons based on water requirement norms.

Water Availability

It is necessary to ascertain the availability position of water from sources like river, canal, ponds, tanks or any other surface water body.

It should preferably be perennial in nature and should have adequate flow to support the quantity of water to be pumped for the proposed scheme.

Water Lifting Permission

Water lifting permission should be obtained for private individuals or societies from Government Department authorized to issue such permissions. The water lifting permission should indicate the period, area and percentage of different crops in each season.

Long duration water lifting permission is required if the implementing agency desires to take loan.

Soil Characteristics

Soil characteristic should broadly be known for its suitability for various crops proposed in the scheme.

Power

Normally the power availability for large schemes is considered to be 16 hours per day but in case of small lift irrigation scheme up to 40 ha command area, power availability may be considered at maximum of 12 hours.

If the command area is less than 10 ha then 8 hours of power supply can be considered on economic considerations.

Intake Well

This is a civil structure required for guiding the water in the sump well / jacks well. In some cases, this structure is necessary even for small schemes to take care of water level fluctuations in the river. It also provides silt free water for the pumping operations. However, for small lift schemes, this need not be insisted upon.

Suction Pipes for Pump set

The following table may be followed in determining the diameter of suction and delivery pipes for various discharge ranges:

Discharge (lps)	Diameter of Suction pipe (mm)
5	65
10	80
15	100
20	125
30	150
50	200
100	300 (NABARD)

Rising Main

Rising main should be designed based on discharge and total pumping head. It is expected that the total pumping head in normal cases will not exceed 50m. The total length of the rising main should not exceed 3000 m.

The diameter of rising main recommended for RCC pipes and Rigid PVC pipes for various discharges are given below:

(a) RCC Pipes

Discharge (lps)	Dia (mm)	Discharge (lps)*	Dia (mm)
5	100	55	300
10	150	60	300
15	200	65	300
20	200	70	300
25	225	75	300
30	225	80	300
35	250	85	350
40	250	90	350
45	250	95	350
50	250	100	350

*(lps = litre per second)

(b) Rigid PVC Pipes

Discharge (lps)	Dia (mm)	Discharge (lps)	Dia (mm)
5	90	45	225
10	125	50	250
15	150	55	250
20	150	60	250
25	200	65	250
30	200	70	300
35	200	75	300
40	225	80	300

If direct tapping on the rising main is proposed for drawing partial discharge then such tapping should not exceed 2 numbers.

Water hammer control devices in case of large schemes are necessary but in case of small lift irrigation schemes, it may not be required. However, the accessories like air valve, drain valve may be provided as per the topographic conditions.

Apart from the PVC and RCC pipes, HDPE and AC pipes of equivalent class and diameter can also be allowed.

Cast Iron (CI) and Mild Steel (MS) pipes are not recommended except for road crossing and nala crossing.

Delivery Chambers

The delivery chamber of one minute retention capacity can be provided for release of water in the command area.

Distribution System

Generally a chak size of 8 ha may be considered while designing the distribution system.

If scheme economics does not permit use of underground pipes then open channels can be provided. For schemes up to 8 ha. command area only open channels need be provided.

Pumping Machinery

Most of the small lift irrigation schemes require centrifugal pump sets for pumping surface water / groundwater.

In case the fluctuations between low water level and high flood level is less than 4.5 m, then the puppets can be placed in the pump house subject to suitability of site conditions.

If the pump sets are to be operated with shifting arrangements then it should not exceed 20 HP per unit.

The permanent pump house can be provided if the water level in the river is more or less constant throughout the year. This may vary from site to site depending upon the stream hydrology.

The centrifugal pump sets selected should conform to BIS standard i.e., IS:10804-94 for complete pumping system.

The usual formulae that may be applied for calculating the horse power of the centrifugal pump sets and discharge required to be pumped are given as under:

1. Break Horse Power

$$BHP = (Q \times H) / (75 \times e)$$

Where, BHP = Break Horse Power of the Centrifugal Pump set

Q = Discharge in liters per second

H = Total Head in meters (including friction losses)

e = Overall efficiency of Pump set (as percentage)

2. Discharge required to be pumped:

The discharge required for the envisaged command area may be calculated from the following formula: $Qr = (28 \times A \times I) / (R \times t)$

Where, Qr = Required discharge in litres per second (lps)

A = Crop area in ha

I = Depth of irrigation in cm

R = Rotation period in days

t = Working hours per day

The discharge required to be pumped during different crop seasons can be calculated for kharif, rabi and summer seasons and the peak discharge i.e., maximum discharge should be adopted as the required discharge (Qr).

Number of Stages

From the management point of view, the number of stages should not exceed two. Especially small lift irrigation projects, as far as possible, there should be only single stage pumping.

Project Cost

The project cost of the scheme should be based on the schedule of rates prevailing in the area.

Command Area Map

A map of the command area showing layout drawing of civil works should be attached to the scheme along with a longitudinal section (L Section) for rising main indicating the Low Water Level (LWL), High Flood Level (HFL) position of fixed foundations, pump set, etc.

Small Lift Irrigation Design on Surface Water

A typical design of small lift irrigation scheme on surface water body is given below.

Basic details of the lift irrigation scheme	
Details	Quantity
Command Area (Ha) =	20
Cropping Pattern (ha)	
Kharif:	
Paddy =	10
Makai=	10
Rabi:	
Wheat	8
Gram	5
Makai	2
Summer	
Makai	5
Static Head (m) =	50
Length of rising main (m) =	3000
No of pumping hours (hrs)	12
The required design is as under:	
Max discharge (cum/hr=	52.778
Total Head (m)	69.68
Hp of pump sets =	27.24
No of pump sets=	2
Diameter of rising main (m)	0.143
Class of pipes = varies from 2.5 kg to 8 kg of different lengths.	

The design varies depending on various factors like, command area, static head, length of rising main, cropping pattern etc. The actual design should be based on the field conditions

Groundwater Based Small Lift Irrigation Schemes

If the lift irrigation scheme is proposed on the sub surface flow in a given alluvial/ hard rock formation then the intake well has to be designed as per the design criteria adopted for a dug well determined on the basis of aquifer characteristics.

If the source of water is an existing irrigation well, the well discharge and command area to be adopted under small lift irrigation scheme could be similar to one that is as usually considered under a dugwell scheme. For schemes, contemplating well discharge of more than 6 lps in hard rock and 10 lps in alluvial areas, the availability of groundwater from such dug wells has to be ascertained from local groundwater department including stipulations of spacing criteria and clearance of scheme based on categorization of blocks/watersheds etc.

Tube Well Irrigation

Tube well irrigation comes under lift irrigation. Tube well is a small diameter hole drilled in the subsoil formation. The cross-sectional area of this type of well is small and unless some mechanical power is used for lifting water, rate of water withdrawal will be low.

The velocity of water withdrawal should be more to maintain high discharge. Obviously the limit of critical velocity is exceeded. Under such conditions the soil particles are dislodged. To prevent clogging of the bore hole a metal pipe with perforations is driven in the hole. It prevents the soil particles from entering the hole. To increase the efficiency of the tube well, in the space between perforated metal tube and soil some straining device is always provided.

The perforations in the metal pipe are necessary only for the depth of water bearing subsoil formations. Other portion of the pipe is generally kept plain. Even though the perforations prevent average soil particles from entering the hole, finer particles enter the hole. To keep the hole clean to avoid settling of particles it is essential to keep the velocity higher (say not less than 1 meter/sec).

The tube well may be taken down to 50 meters in the water bearing stratum. When mechanical power is used to lift water average discharge of tube well is 0.04 m3/sec. The factors which affect the location of tube well are more or less, the same as those discussed for open wells.

Tube well irrigation is very suitable where the subsoil formation is suitable for storing water. Water bearing layer may be of unlimited extent (unconfined) or it may be alternate to some type of impervious layers or may be confined between two impervious layers. In Haryana, Uttar Pradesh, Punjab and Bihar States tube well irrigation is extensively in practice. The obvious reason is that most of the area is made up of alluvium which holds the water to a great extent and there are massive groundwater reservoirs.

Types of Wells

The utilization of groundwater through dug well irrigation is an indigenous form of irrigation. A dug well is a shallow well, with its bottom on a fair depth below the water table, so that water from the surrounding aquifer accumulates in the well. Water collected in the well is lifted to ground surface through a water lift. The masonry lined dug well usually yields 7 to 8 m3 per hour (@ 2 1ps) when operated with a Persian wheel, which is the case for about 20% of the masonry wells. In the remaining dug wells, the water is lifted by animal power with leather or metal buckets, usually bullocks. These wells have very limited discharge rates and this practice is almost dispensed with due to high cost of labour and drudgery.

- Shallow tube wells are drilled to penetrate a shallow aquifer and are usually less than 30 m deep. This depth is only possible when the tube well is placed at the bottom of a dug well, so that it is a dug-cum-tube well. Shallow tube wells are usually equipped with a small centrifugal pump. The filter point wells if Cauvery deltas of Tamil Nadu in Trichy, Thanjavur, Thiruvarur, Nagapatinam districts are shallow tube wells irrigating individual farms. The electric or diesel motor is directly connected to the pump by a belt drive. The centrifugal pump is placed at the surface level and operates mainly in suction mode. These wells usually have a capacity of 20 to 30 m3h-1 (@ 7 lps).

- Medium tube wells are small diameter submersible tube wells equipped with a strainer section. These wells are usually about 45 m in depth although they may be deeper depending on the depth of the aquifer and the capacity desired. They usually have capacities of about 30 to 40 m3h-1 (@ 10 1ps) and are equipped with centrifugal pumps. Water distribution from these wells is through small unlined channels with the following lengths: masonry wells - 30 m; shallow tube wells - 200 m; and medium tube wells - 400 m.

- Deep tube wells have a large diameter and vary in depth from 40 to 300 m. Pumps are sunk into the well, operate in force mode, and are driven by submersible electric engines or by shafts connected to engines at the surface. Deep tube wells have a large discharge capacity varying from 150 to 300 m3h-1 (40 to 80 lps). As discharge capacity increases, the length of the water distribution channels increases accordingly. For example a command area of 100 ha is served by water distribution system of 4 km in state tube well commands of UP, water is distributed through unlined earthen channels.

State Tube Well (STW) Programs

A major instrument of public policy – State Tube well (STW) programs – is devised originally to stimulate groundwater irrigation and to ensure that the access to this communal resource is diffused and is not monopolized by the rural elite. For hard rock regions, open dug wells are technically found to be ideally suited. Average command area of dug wells being rather low (less than 2 hectare), it would mean, in operational terms, government coping with an innumerable number of open dug wells. The other major problem with public tube well programs is their management, efficiency and quality of irrigation service they are able to provide. Numerous field studies, have pointed out poor maintenance, lack of accountability of the tube well operator of the community, domination by local elite, frequent power cuts, delays in repair and procurement of spare-parts, local feuds regarding the right of passage, etc., are amongst the several problems that STW programs suffer from. It was found that small farmers did benefit from public tube wells through improvement in crop pattern, crop yields, and cropping intensity. However, overall experience with public tube wells in various regions of the country is quite disappointing from the point of efficiency. As far as equitable distribution is concerned, the rural elite usually succeed in appropriating most of the benefits of public supply.

Merits of Well and Tube Well Irrigation

1. Well is simplest and cheapest source of irrigation and the poor Indian farmer can easily afford it.

2   Well is an independent source of irrigation and can be used as and when the necessity arises. Canal irrigation, on the other hand, is controlled by other agencies and cannot be used at will.

3. Excessive irrigation by canal leads to the problem of reh which is not the case with well irrigation.

4. There is a limit to the extent of canal irrigation beyond the tail end of the canal while a well can be dug at any convenient place.

5   Several chemicals such as nitrate, chloride, sulphate, etc. are generally found mixed in well water. They add to the fertility of soil when they reach the agricultural field along with well water.

6. The farmer has to pay regularly for canal irrigation which is not the case with well irrigation.

Demerits of Well and Tube Well Irrigation

1. Only limited area can be irrigated. Normally, a well can irrigate 1 to 8 hectares of land.

2. The well may dry up and may be rendered useless for irrigation if excessive water is taken out.

3. In the event of a drought, the ground water level falls and enough water is not available in the well when it is needed the most.

4. Tubewells can draw a lot of groundwater from its neighbouring areas and make the ground dry and unfit for agriculture.

5. Well and tube well irrigation is not possible in areas of brackish groundwater.

Sewage Irrigation

Sewage water is used as potential source irrigation for raising vegetables and fodder crops around the sewage treatment sites, which are directly or indirectly consumed by human beings. Raw Sewage is a rich source of organic and inorganic nutrients for plant growth; sewage farming is quite common in all urban areas in India. Some cities sewage water where industrial effluent is discharged along with may contain toxic metals in high amounts. Thus the composition of domestic sewage may vary with the type of industrial discharge their waste. The analysis of eight chemical parameters of untreated sewage water was carried out in Vidyaranyapuram STP, Mysore, India. Factor analysis was applied on untreated sewage water. Data matrix, pollution factor was found to be the most contributing factor and it showed 22.31% of the total variance (Chloride, BOD, COD and TDS). The second most contributing factor was found to be the nitrification factor, which explained 21.11% of the total variance (pH and Nitrate), whereas the salinization factor contributed 16.98% of the total variance (TS and TSS). Sewage has high values of temperature, pH, hardness, alkalinity, chemical oxygen demand, total soluble salts, nitrates, nitrites and cations like sodium, potassium, calcium and magnesium.

The sewage water has a high fertility load; it adds available N, P, K, Fe, Mn, Zn and Cu to soil, this indicating their significant addition through sewage suggesting use of sewage water as a low grade cheap fertilizer in agriculture. It can drastically reduce the cost due to substitution of chemical fertilizers. The concentration of heavy metals such as Cu, Pb and Co in plant tissue was low compared to limits standards. These heavy metal concentrations are well below hazardous levels. The effect of continuous irrigation with sewage water increases exchangeable cations to a large extent. Sewage water application increases the soil salinity, organic carbon, N, K, Ca, Mg cations to a lot. Soil is a biofilter that can reduce a large part of domestic waste water pollutants, but this filtering increase EC, SAR, Na, Ca and Mg of soil. In addition to these, sewage water also contains significant amounts of toxic metals such as arsenic, chromium, cadmium, copper, lead, nickel, zinc, cobalt, magnesium and iron.

Sewage Farm

Sewage Farm is a plot of land used for the natural biological decontamination of sewage liquids and for raising agricultural crops. In the USSR a distinction is made between municipal and agricultural farms. The former are established on land managed by municipal agencies. They basically perform a sanitary function and differ little in layout from leach fields. Agricultural sewage farms are set up on kolkhozes and sovkhozes for raising crops that use the nutrients contained in sewage liquids. Some are used only in the summer, and others operate year-round.

The sewage farms followed a dewatering concept by J. Hobrecht. In 1869 the Berlin administration made him director of the Berlin Latrine System. Hobrecht divided the city into 12 districts, called radial systems. Each radial system had a pumping station. Pumping stations received domestic, commercial and industrial waste waters as well as precipitation water through gravity flow pipelines. Sewage effluents were conducted from the pumping station through pressure pipelines to sewage farms located outside the city. Some sewage farms were additionally supplied by direct pipelines.

Pressure pipelines discharge waste water at the sewage farms. Waste water is first collected in sedimentation basins made of concrete or earth. Water flows through the tank and most sediments settle to the bottom. Immersion panels hold back floating matter. Sediments settling in the sedimentation basin are regularly evacuated and dewatered at special sludge drying areas. Dewatered sludge was used as a soil conditioner for agriculture and horticulture in early years. The sewage farm trench system is also regularly cleaned, whereby removed sediments are usually deposited directly alongside the trench. After sewage water has passed through the sedimentation basin, e.g. has been mechanically cleaned, it flows through gravity feeders to the terraces.

The natural ground form was not automatically suited for processing sewage waters. Terraces were constructed horizontally or sloping, depending on the surface. They were about 0.25 ha large, and surrounded by embankments. There are three methods of sewage farm treatment. Horizontal terraces are flooded by surrounding distribution ditches. For slope terraces, sewage water overflows the upper bank and irrigates the sloped terrace. Bed terraces with ditch irrigation were also initially used. Waste water flowed through bed terraces in connected parallel furrows, about a meter apart. Only plant roots received water.

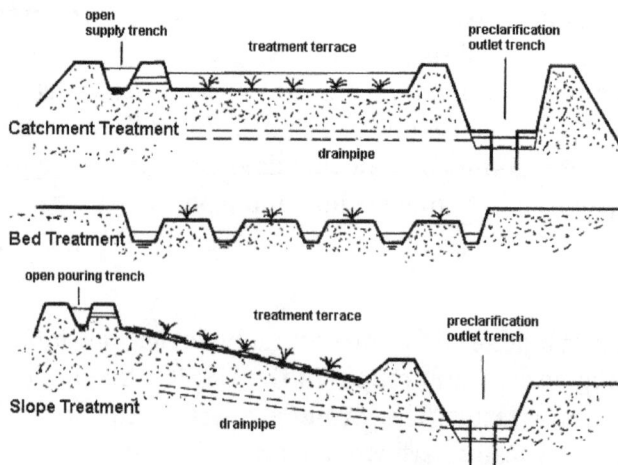

Types of sewage farms

Wild sewage areas are often found near treatment terraces. The overloading of prepared surfaces can be met by directly diverting unpurified water through sluices onto natural land. Sewage water contents are retained during the passage through the soil, adsorbed in topsoil without humus, and handled chemically and biologically. This process supplies agriculturally useful nutrients. Initial yields were high and the majority of fields were used agriculturally and served their own sewage treatment plots. There was a mixed use of grasslands and field cultivation.

Most sewage farms were provided during construction with drainage pipelines at regular intervals for a faster discharge of filtered and purified water, and to provide for aeration and regeneration of soils as well. Drainage water passes through collecting drains and dewatering trenches into the preclarification outlet trenches. Some water from soil passage percolates into ground water.

Fields are flooded in normal operation, and then left until water seeps away and the soil is re-aerated. The next flooding is begun only after re-aeration is completed. These sewage farm rhythms are also oriented to the growth periods of agricultural crops. Four to eight field treatment cycles a year are possible on grasslands, with 2,000-4,000 mm of sewage water. Areas used for cultivation of winter wheat can only be used once a year, with 100-500 mm of waste water.

Sewage farms were overtaxed with increasing amounts of waste water, intensification of agricultural production, and the closure of other sewage farms. This stimulated some sewage farm operators to establish "intensive filter areas". These are permanently flooded and surrounded by high embankments. An inadequate degree of purification is performed here because aerobic processes cannot take place. These areas were not used agriculturally.

Sewage farm structures were often leveled after sewage treatment use was discontinued. Trenches and terraces were filled with material from the embankments, themselves land-fill material.

Advantage

Sewage farming allows use for irrigation of water, which might otherwise be wasted. Some of the nutrients and organic solids in wastewater can be usefully incorporated into soil and agricultural products rather than fouling natural aquatic environments. Pumping to the point of application may be the only requirement if the village is not at a higher elevation than the sewage farm.

Disadvantages

Polluted runoff may occur from sewage irrigation of fields when entering wastewater and precipitation exceed evaporation and percolation capacity.

Sewage is usually generated at a relatively constant rate, but irrigation is required only during dry weather, and is useful only while temperatures are high enough to promote plant growth. Over-irrigation causes soils to become septic, sour, or *sewage-sick*. Arid climates may allow temporary storage of sewage in holding ponds while the soils dry out during non-growing seasons, but such storage may cause odor and aquatic insect problems, including mosquitoes.

It may be impractical to protect the crops being grown from sewage contact. Even optimum situations like irrigating fruit trees with flow in surface ditches may involve some risk of pathogen transfer from the sewage to the edible fruit by birds, insects, and similar vectors. Pathogen transfer is more likely with ground crops, and practically unavoidable with spray irrigation.

Similar Wastewater Systems

Subsurface distribution piping is problematic since it is vulnerable to root blockage and to damage during soil cultivation. Also obstruction of distribution piping by sewage solids discourages sewage farming when wastewater is not pre-treated as it is typically the case in a septic drain field.

Sewage farming should not be confused with wastewater disposal through infiltration basins or subsurface drains.

Plows or harrows may be used to periodically break up vegetation mats which are slowing surface disposal.

Subsurface disposal typically uses pipes placed deep enough to minimize root penetration and often manages overlying vegetation to avoid growth of plants with deep root systems.

Supplemental Irrigation

Supplemental irrigation (SI) can be defined as the addition to essentially rain-fed crops of small amounts of water during times when rainfall fails to pro-vide sufficient moisture for normal plant growth, in order to improve and stabilize yields. Accordingly, the concept of SI in areas with limited water resources is based on the following three aspects.

Surface water sources are the cheapest supply for supplemental irrigation

First, water is applied to a rain-fed crop, that would normally produce some yield without irrigation.

Second, since precipitation is the principal source of moisture for rain-fed crops, SI is only applied when precipitation fails to provide essential moisture for improved and stabilized production.

Third, the amount and timing of SI are scheduled not to provide moisture stress free conditions throughout the growing season, but to insure that there is a minimum amount of water available during the critical stages of crop growth that would permit optimal instead of maximum yield.

In many areas, groundwater is being over-exploited and sustainability threatened

The management of supplemental irrigation is seen as a reverse of that of full or conventional irrigation (FI). In the latter, the principal source of moisture is the fully controlled irrigation water, while the highly variable limited precipitation is only supplementary. Unlike FI, the management of SI is dependent on the precipitation as a basic source of water for the crop.

Water-harvesting is a sustainable source of supple-mental irrigation

Water resources for supplemental irrigation are mainly surface, but shallow groundwater aquifers are being increasingly used. Of the non-conventional water resources, such as treated sewage effluent, which have potential for the future, water harvesting is an important one.

Improving Production with Supplemental Irrigation

Research results from ICARDA and other institutions in the dry areas, as well as data from farmers' fields, show substantial increases in crop yields in response to the application of relatively small amounts of SI. This increase is possible in both low and high rainfall conditions. Average increases in wheat grain yield under low, medium and high annual rainfall at Tel Hadya were about 400%, 150% and 30% using SI amounts of about 180, 125 and 75 mm, respectively. Generally, optimal SI ranges from 75 mm in areas with annual rainfall of 500 mm, to 250 mm in areas receiving 250 mm rain. Determining the optimal amount of SI under various conditions is discussed later.

When rainfall is low, more water is needed, but the response is greater; however, increases in yield are remark-able even when rainfall is as high as 500mm. The response is greater when rain-fall distribution over the season is poor. However, in all rain-fed areas of the WANA region, some time in the spring there is usually a period of stress, which threatens yield levels. Soil-moisture stress usually starts in March, April or May if total seasonal rainfall is low, average or high, respectively.

Yield is much higher with supplemental irrigation

In Syria, average wheat yields under rain-fed conditions are only 1.25 t/ha and this is one of the highest in the region. With SI, the average grain yield rises to 3 t/ha. In 1996, over 40% of rain-fed areas were under SI and over half of the 4 million tonnes of national production was attributed to this practice. Supplemental irrigation not onlyincreases yield, but also stabilizes production. The coefficient of variation in production in Syria was reduced from100% in rain-fed crop to 10% when SI was practiced. This stability is of great importance since it secures farmers' income.

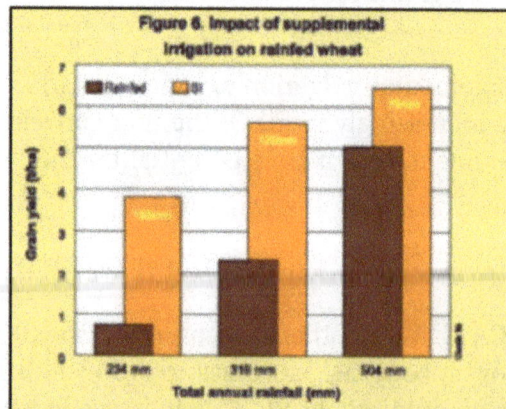

High Water-use Efficiency

Water-use Efficiency (WUE) is a measure of the productivity of the water consumed by the crop. In areas with limited water resources, where water is the greatest limitation to production, WUE is the main criterion for evaluating the performance of agricultural production systems. No longer is productivity per unit area the main objective, since land is not as limiting to production as is water.

Average WUE of rain in producing wheat in the dry areas of WANA is about 0.35 kg grain/m3, although with good management and favorable rainfall(amount and distribution), this can be increased to 1 kg grain/m3. However, water used in SI can be much more efficient. Research at ICARDA showed that a cubic meter of water applied at the right time (when the crop is suffering from moisture stress), combined with good management, could produce more than 2.5 kg of grain over the rain-fed production. This extremely high WUE is mainly attributed to the effectiveness of a small amount of water in alleviating severe moisture stress during the most sensitive stage of crop growth and seed filling. When SI water is applied before such conditions occur, the plant may reach its high yield potential.

In comparison to the productivity of water in fully irrigated areas (when rain-fall effect is negligible), the productivity is higher with SI. In fully irrigated areas with good management, wheat grain yield is about 6 t/ha using 800 mm of water. Thus, the WUE is about 0.75kg/m3, one-third of that under SI with similar management. This suggests that water resources may be better allocated to SI when other physical and economic conditions are favorable.

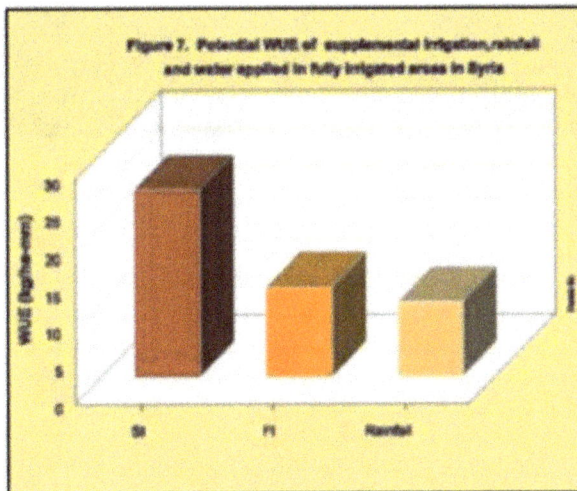

Figure 7. Potential WUE of supplemental irrigation, rainfall and water applied in fully irrigated areas in Syria

Management of Supplemental Irrigation

The most important considerations in good SI management are when and how much water to apply. Yet many, perhaps most, farmers apply too much water if they can get it. Evidence of the over-use of irrigation water is clear in many dry-area situations, and SI is no exception. Farmers tend to overuse water in SI because of the low water and irrigation costs. The aim of any management pro-gram for SI should be to provide sufficient water to crops at the right time and also to discourage farmers from over-irrigating.

When to use Supplemental Irrigation

Unlike in full irrigation, the time of SI irrigation cannot be determined in advance. This is due to the fact that the basic source of water to rain-fed crops is rainfall, which being variable in amount and distribution, is difficult to predict. Since SI water is best given when soil moisture drops to a critical level, the time of irrigation can be best determined by measuring soil moisture on a regular basis. Unfortunately, there is no simple device that an average farmer can use for this purpose. The well-known tensiometers are not suitable, since SI allows lower soil-moisture potential than the tension meter is able to read correctly; other more sophisticated methods are not appropriate either.

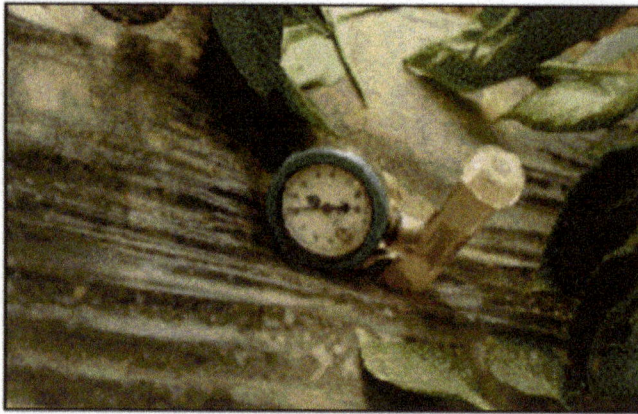

Instead, most farmers in the region rely on personal experience related to the amount of rainfall received and crop appearance. Generally, they tend to irrigate earlier than necessary with more frequent irrigation than needed when they have on-demand water supply.

Cultural Practices

Soil Fertility

SI alone, although it alleviates moisture stress, cannot ensure good performance of the rain-fed agricultural sys-tem. It has to be combined with other farm management practices. Of most importance is the soil fertility; particularly in the Mediterranean region nitro-gen is usually the main deficiency. Absence of nutrient deficiency greatly improves yield and water-use efficiency.

Nitrogen deficiency affects both yield and water-use efficiency.

Under rain-fed conditions, the rate of fertilizer nitrogen needed is not high, and with some water stress, high rates may actually be harmful. Under the Syrian rain-fed conditions, 50 kg N/ha was sufficient. However, with more water applied, the crop responds to nitrogen up to 100 kg N/ha, after which no benefit is obtained. This rate of N greatly improves WUE. It is also important that there is adequate avail-able phosphorus in the soil so that response to N and applied irrigation is not constrained. Other areas may have deficiencies in other elements. It is always important to eliminate these deficiencies to increase yield and WUE.

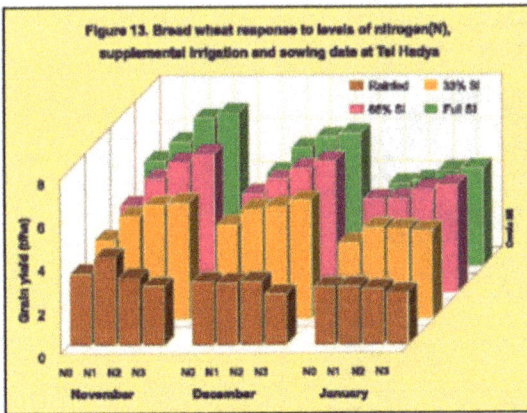

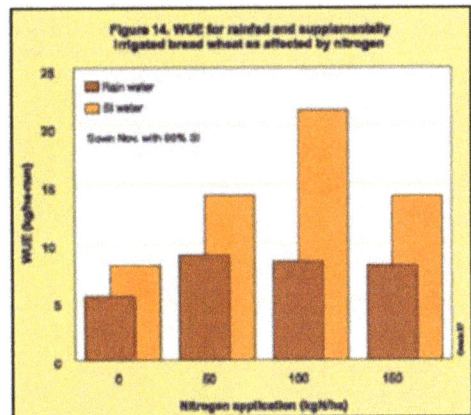

Crop Varieties

Selection of the appropriate crop variety makes a difference under both rain-fed and SI conditions. In rain-fed areas, breeding seeks to produce drought resistant varieties. These perform well under rain-fed conditions, but because they were not developed for SI conditions their response to more reliable water supplies may not be high. A suitable variety for SI is one with good response to limited water application while maintaining some drought resistance.

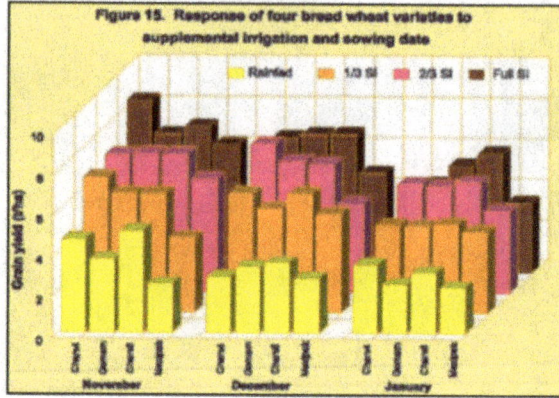

Figure 15. Response of four bread wheat varieties to supplemental irrigation and sowing date

Planting Date

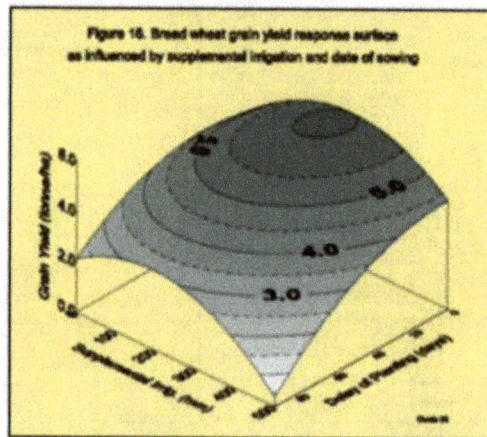

Figure 16. Bread wheat grain yield response surface as influenced by supplemental irrigation and date of sowing

Under rain-fed farming, the earliest planting is usually in November, when enough rain for germination has fallen. Around 15 November is the optimal date for achieving highest yield under rain-fed conditions in the eastern Mediterranean region. Delaying planting after this date has negative impact on yield. Under SI, early planting (1 November) has an advantage in both yield and WUE. With recommended levels of SI and N, wheat yield was decreased substantially by delaying sowing from December to January; for WUE the decrease occurred mainly with the delay from November to December with little decrease afterward. With SI, early planting is possible unlike under rain-fed conditions, where one has to wait for enough rain. Delaying planting, however, is not always a disadvantage for SI. Planting in mid-December and mid-January delayed flowering and grain-filling by one and two weeks, respectively, compared with the crop planted in early November. This shift delays the need for SI, and a system of phased planting may be used to spread the peak water requirements during spring. This would reduce the need for high water supply discharge and reduce the size of the irrigation system needed. Knowing that SI cost is a major issue, this concept can help to make it more economic.

Irrigation Systems

The difference between SI and full irrigation is in the management, not in the physical system. Irrigation systems that are suitable for full irrigation are also technically suitable for SI. There are also economic considerations. Systems for SI are only occasionally used. In most of the rain-fed areas only one to three irrigations may be needed each year. Using a system occasionally rarely justifies large investment. This may be why most SI systems are of the surface type, although they have lower application efficiency and uniformity. More costly systems, such as sprinklers and trickle are mainly acquired for full irrigation in summer but utilized for SI in winter. When labor and water are costly and field is not suitable for surface system, or no full irrigation systems exist, then sprinkler systems are generally used. The objective then is to minimize the cost by appropriate selection of the type and size of the system. One problem can be that the need for water occurs at the same time for all the planted area. Irrigation has to be applied to a large area in a short time. A large water sup-ply rate and a large irrigation system are then needed. This conflicts with the objective of minimizing the cost. To over come this problem the following strategies may be adopted.

1. The use of mobile sprinkler systems that can be moved easily within the field either mechanically or by hand when labor is cheap. Hand-moved systems are suitable when labor is not very costly. However, side-roll and gun sprinklers also give good value for money.

2. The problem of the requirement of large discharge rate and system size during the narrow peak period can be reduced substantially by spreading the peak water requirements over a longer period in late spring. This can be done by using different planting dates in different parts of the field, ranging from early November to late January, and selecting different crops and varieties that require water at different times. Although this may affect the yield and production a little, it will cut the cost of irrigation.

3. During the peak irrigation period, the amount of water required may be divided into two or more applications. A smaller amount is given, but covering the field faster avoids severe stress.

4. Water can be stored in the soil profile before the beginning of the irrigation time. During the early spring, crop consumption rate increases and some soil storage space is created before the critical level is reached. During this time farmers can start irrigating early and fill part of this space, which in turn will delay the time by which next irrigation will be needed. This will educe the peak water need. In fact, some farmers who have small irrigation systems (one hand-moved system) already practice this.

5. When the farm has irrigated summer crops, the system must be designed to suit both full and supplemental irrigation.

6. When the system is only for SI, size and cost may be reduced by considering the optimal water scheduling needs, not the crop's full water requirement. Providing 50% of the full irrigation requirements not only increases WUE and net return, it also cuts water need, irrigation system size and cost.

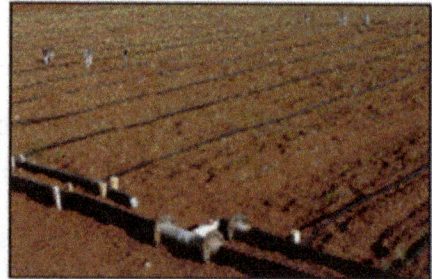

Hand-moved sprinkler system is low in cost butrequires extensive labor.

Attempts to adapt furrow irrigation for supplementalirrigation of wheat
may help in reducing irrigationcost while maintaining high efficiency

Surface irrigation is very popular. It is labor-inten-sive but requires no capital investment.

References

- Reed, Sherwood C.; Middlebrooks, E. Joe; Crites, Ronald W. (1988). Natural Systems for Waste Management & Treatment. McGraw-Hill. pp. 220&246. ISBN 0-07-051521-2.

- Soil-and-organics/what-is-gravity-irrigation, growology: gardenculturemagazine.com, Retrieved 10 May 2018

- "Welsh History Review - Vol. 14, nos. 1-4 1988-89 Merthyr Tydfil in the mid-Nineteenth Century: the struggle for public health". The National Library of Wales, Aberystwyth. Retrieved 23 October 2016.

- Wells-and-tube-wells-irrigation-in-india-merit-and-demerits-21109: yourarticlelibrary.com, Retrieved 26 June 2018

- David Williams (1991). "The rehabilitation of the River Taff". National Rivers Authority. Retrieved 29 October 2016.

- Supplemental-Irrigation-A-Highly-Efficient-Water-Use-Practice-267138098: researchgate.net, Retrieved 21 April 2018

Permissions

All chapters in this book are published with permission under the Creative Commons Attribution Share Alike License or equivalent. Every chapter published in this book has been scrutinized by our experts. Their significance has been extensively debated. The topics covered herein carry significant information for a comprehensive understanding. They may even be implemented as practical applications or may be referred to as a beginning point for further studies.

We would like to thank the editorial team for lending their expertise to make the book truly unique. They have played a crucial role in the development of this book. Without their invaluable contributions this book wouldn't have been possible. They have made vital efforts to compile up to date information on the varied aspects of this subject to make this book a valuable addition to the collection of many professionals and students.

This book was conceptualized with the vision of imparting up-to-date and integrated information in this field. To ensure the same, a matchless editorial board was set up. Every individual on the board went through rigorous rounds of assessment to prove their worth. After which they invested a large part of their time researching and compiling the most relevant data for our readers.

The editorial board has been involved in producing this book since its inception. They have spent rigorous hours researching and exploring the diverse topics which have resulted in the successful publishing of this book. They have passed on their knowledge of decades through this book. To expedite this challenging task, the publisher supported the team at every step. A small team of assistant editors was also appointed to further simplify the editing procedure and attain best results for the readers.

Apart from the editorial board, the designing team has also invested a significant amount of their time in understanding the subject and creating the most relevant covers. They scrutinized every image to scout for the most suitable representation of the subject and create an appropriate cover for the book.

The publishing team has been an ardent support to the editorial, designing and production team. Their endless efforts to recruit the best for this project, has resulted in the accomplishment of this book. They are a veteran in the field of academics and their pool of knowledge is as vast as their experience in printing. Their expertise and guidance has proved useful at every step. Their uncompromising quality standards have made this book an exceptional effort. Their encouragement from time to time has been an inspiration for everyone.

The publisher and the editorial board hope that this book will prove to be a valuable piece of knowledge for students, practitioners and scholars across the globe.

Index